Regionalism in the European Union

Edited by
Peter Wagstaff

intellect™

EXETER, ENGLAND
PORTLAND (OREGON) USA

intellect – EUROPEAN STUDIES SERIES

General Editor – Keith Cameron

Humour and History	Keith Cameron (ed)
The Nation: Myth or Reality?	Keith Cameron (ed)
Regionalism in Europe	Peter Wagstaff (ed)
Women in European Theatre	Elizabeth Woodrough (ed)
Children and Propaganda	Judith Proud
The New Russia	Michael Pursglove (ed)
English Language in Europe	Reinhard Hartmann (ed)
Food in European Literature	John Wilkins (ed)
Theatre and Europe	Christopher McCullough
European Identity in Cinema	Wendy Everett (ed)
Television in Europe	James A. Coleman & Brigitte Rollet (eds)
Language, Community and the State	Dennis Ager
Women Voice Men	Maya Slater (ed)
National Identity	Keith Cameron (ed)
Policing in Europe	Bill Tupman & Alison Tupman
Regionalism in the European Union	Peter Wagstaff (ed)
Spaces in European Cinema	Myrto Konstantarakos (ed)

First Published in United Kingdom 1999 by
Intellect Books
School of Art and Design, Earl Richards Road North, Exeter EX2 6AS

First Published in USA 1999 by
Intellect Books
ISBS, 5804 N.E. Hassalo St, Portland, Oregon 97213-3644

Series Editor: Keith Cameron
Copy Editor: Lucy Kind
Cover Illustration: Mark Mackie

A catalogue record for this book is available from the British Library

ISBN 1-84150-001-1

Printed and bound in Great Britain by Cromwell Press, Wiltshire

Contents

Foreword

Since I became an MEP in 1979 an understanding of the regions of Europe has become increasingly central to an understanding of the European Union as a whole. Now, as Minister for Europe, I frequently have the pleasure of working with regional representatives.

As the impact of European Community decision-making on people's lives has increased, so the regions have increasingly found a voice, through the Committee of the Regions and within Member states. This enables them to influence those decisions and benefit from the opportunities which the European Union has to offer the regions. In the United Kingdom, devolution will further boost their role.

I know that *Regionalism in the European Union* will help all those interested in Euopean Politics and the European Community to appreciate the collective impact of regions at a European level, and to understand the differing role they play within each Member state.

Joyce Quin, M.P.,
Minister of State, Foreign and Commonwealth Office

Preface

When I wrote the introduction to *Regionalism in Europe* in 1994, just five years had elapsed since the Berlin Wall was pulled down. The assertion I made then – that the pace of change in Europe had made the task of predicting future developments extremely hazardous – seems no less valid today (Wagstaff, 1994, p. 1). A recent commentator has underlined the inconvenient tendency for events to surprise those who would analyse them: 'The almost universal failure to predict the collapse of communism drove a large nail into the coffin of Western political science' (Mazower, 1998, p. 367). Within the European Union, the post-Maastricht progress towards economic and monetary union is approaching completion, or at least the end of the beginning, albeit without the participation in its first phase of one of the European Union's major Member States, and in the face of muted enthusiasm among the citizens of the countries concerned. It may take some time before deeply-felt attachment to the French franc or the German mark, symbols of a security which is more than simply economic, is usurped by affection for the Euro.

The goal of political union, seen by many as the desired concomitant to economic union, is still a rather hazy outline in the distance, though is perhaps closer than before. Further afield, Mikhail Gorbachev's image of 'the common European house' has taken on the patina of age (while the image of Gorbachev himself is fading rapidly), yet the concept still embraces land where the horrors of 'ethnic cleansing' persist — we only need to substitute Kosovo for Bosnia Herzegovina to find the same bloody effects of ethnic nationalism. The only difference is that what was once seen as a grotesque neologism has now, it would seem, slipped unnoticed into everyday language.

There is one conclusion that may be drawn with some confidence from all this. In the realm of regional, national or continental identities and allegiances there are few certainties. What seems permanent is merely a snapshot of an historical moment at which, for example, the apparently solid structure of the nation-state may be only days or weeks away from profound change. If, in spite of all the uncertainties, there is a future to be foretold, it is perhaps still one in which states are confronted with the need to reassess their relationships, both with their neighbours and with their citizens. One in which boundaries, both external and internal, are redrawn and decision-making is no longer the exclusive prerogative of national governments acting alone or in concert with each other. As it happens, the preamble to the Maastricht Treaty contains the declaration that the Member States of the European Community are determined 'to continue the process of creating an ever-closer union among the peoples of Europe, where, in keeping with the principle of subsidiarity, decisions are taken as closely as possible to the citizens'. It is in this context that the growth in regional consciousness and the affirmation of regional identities can perhaps best be seen.

Perhaps the most surprising development in the last few years has been the transformation of the regional theme in the one state – the United Kingdom – that, earlier, had seemed the most unsympathetic to a re-evaluation of the relationship between central government and the regions. The impetus given to the cause of devolution within the United Kingdom by the election victory of the Labour Party in May 1997 and by the rapid process of electoral consultation which took place in the following months in Scotland, Wales and Northern Ireland is a further indication, both of the vitality of the subject, and the unpredictability of its development.

The chapters that follow explore and analyse the regional theme throughout the European Union as it is presently constituted. Only Luxembourg has been omitted, as its size precludes sensible comparisons with the other Member States. From Ireland to Greece, from Portugal to Scandinavia, there is an awareness that regions, however variously they are defined, have a contribution to make to the functioning of the nation state and to the European Union as a whole. The extent of these contributions and the readiness of the nation state to foster and encourage them vary greatly, and this, more often than not, is explained by the historical relationship in each case between centre and periphery. An acknowledgement of the importance of the regional theme, and of the pressure for its expression at an institutional level, can be seen in the establishment, by the Maastricht Treaty in 1991, of the Committee of the Regions of the European Union and of its confirmation and consolidation in the years that have followed. Long before this official sanctioning of the regional theme, however, there has been ample evidence that, in many parts of Europe, regional diversity, rooted deeply in culture, language, shared experience and aspirations, is in the ascendant, even though it may be premature to accept unreservedly the claim that 'the ideal of cultural unity in one country - everyone studying the same history book under the portrait of the same President - belongs to the dying ideology of the modern nation-state' (Ascherson, 1988, p. 219).

I would like to thank friends and colleagues, both at Bath and elsewhere, for their helpful comments and their willingness to contribute to this updated, augmented and more complete study of *Regionalism in the European Union*. I am grateful, in particular, to Keith Cameron (General Editor of the Intellect European Studies series), for his patience and, once again, to Hilary Strickland for the preparation of the maps.

Peter Wagstaff
Bath, August 1998

INTRODUCTION:
Regions, Nations, Identities

Peter Wagstaff

What does it mean to be English, or Scottish, or Welsh? Or British? What does it mean to be French, or Breton, or Basque; Spanish, or Catalan, or Basque? What does it mean to be European? If a Scot claims to be Scottish, British and European, how strong are these allegiances, and why? These are questions which hint at the difference between nation and state, but also between the state and something bigger – and smaller. The changing map of Europe over the last decade has focused attention on the question of allegiance, with results that have sometimes exacted, and continue to exact, great cost in human suffering. The phrase 'blood and belonging' (Ignatieff, 1994) encapsulates both the suffering and the visceral attachment to land, roots, and ancestry, an attachment visible in perhaps its rawest form following the disintegration of Communism in Eastern Europe. The 15 Member States of the European Union have, for the most part, escaped the worst excesses of political, ethnic or religious strife in recent times, though there can be no scales in which to weigh extremes of violence and its effects, whether in Omagh or Kosovo. Indeed, it was precisely to remove the pretext and opportunity for conflict between its two largest founding members that the European Community was set up in its earliest form. The majority of the current 15 states have, however, been confronted over the years, and rarely more so than at present, with challenges to the integrity of their institutions, their systems of governance, and in some cases their borders, both external and internal.

These political pressures, which in the post-Cold War era have taken the form of centrifugal forces acting against the unity of the nation state, have been accompanied at the same time by the pressures of a global economy which obliges states to respond, as producers and consumers, to market forces largely beyond their control. It can be argued, therefore, that the nation state is under siege, both from the effects of globalisation and from the disaffection of its constituent parts. In short, the argument runs, it is both too big and too small. The progress towards European Monetary Union and the convergence of the economies of Member States represents an attempt to meet the challenge of the global market, and the principle of 'subsidiarity' enshrined in the Maastricht Treaty of 1991 reflects an awareness in some quarters that various aspects of governance are best exercised at levels other than that of the nation state. Attachment to the latter remains very strong, however, as a headline *'VIVE LA NATION!'* in the French Communist Party newspaper, *L'Humanité*, a week before the Maastricht Treaty referendum in France on 20 September 1992, indicates. As governments of the

European Union move, with varying degrees of enthusiasm, towards agreements on economic and monetary union and the gradual erosion of frontiers, their citizens are faced with novel and bewildering challenges to traditional allegiances. Parties and interest groups from across the political spectrum are predictably keen to exploit the uncertainty in the service of a particular cause – the promotion of a particular view. Strange bedfellows are sometimes created as a result. For example, the stout defence of *la nation* mounted by the French Communist Party in the face of what it sees as the federalist drift implicit in the Maastricht debate recalls de Gaulle's vision – if not his expression – of a *Europe des patries,* and finds a curious echo in the recent assertion by the 90 year-old claimant to the French throne that France is in need of a monarchist revival to save it from the worst excesses of a federal Europe (Henri, 1996). Further echoes, too can be found in the reticence with which many United Kingdom Conservatives, among others, contemplate the word 'federal'.

Elsewhere in Europe, of course, and indeed within the European Union itself, federalism is an accomplished fact, and long accepted as a dominant feature in the political and institutional landscape. For Germans in a newly-united Germany the federal system of the *Bundesrepublik,* a meticulously calibrated balance of power between state and region, has proved itself over nearly 50 years in the 11 western *Länder* and is now having to show its resilience and adaptability in confronting the pressing difficulties of the five new *Länder* in the east. Austria, too, has a federal structure, although this dates from the collapse of the Austro-Hungarian Empire in the First World War.

In this respect, though, Germany, together with Austria, is the exception among those countries under consideration here. Switzerland of course, outside the European Union and therefore beyond the scope of this study, is similarly federal in structure and has been since 1848. In France, Italy, the Netherlands, Spain, Portugal, Greece, the United Kingdom and Ireland, the notion of federalism has no place in the political lexicon. The countries of Scandinavia have some historical claim to what is best referred to as inter-state regionalism, rather than federalism. Only in Belgium, with its very specific cleavage between two cultures and two language communities, has the concept found favour, but in an environment which remains unstable.

If federalism is viewed with suspicion, even distaste, in some quarters, there can however be no denying the strength with which regional identity has made itself felt at so many levels across Europe in recent years. True, the collapse of Communism and the eastern bloc has led to the explosion of pent-up regional and ethnic antagonisms, the results of which are daily and horrifically present and visible in the media. Yet even within the relative stability of the EC, the aspirations of regional entities smaller than and contained within existing nation states have been more and more in evidence. In the space of little more than a generation, regional assertiveness has been felt in most of the countries of the EC. Whether in Scotland or Brittany, Corsica or Catalonia, Lombardy, Flanders or the Basque Country, the seamless and integral nature of the nation state has been called into question, as regionalist movements have sought to shake off the more or less oppressive yoke of central control and to stake their claim to varying degrees of autonomy and regional self-expression.

There is, however, a problem of vocabulary. While 'federalism' may perhaps be neatly encapsulated by the definition of 'a state that divides governmental activities between the centre and regional units in such a way that each has the right to make the final decision in at least some fields of activity' (Rokkan and Urwin, 1982, p. 234), when it comes to nations and nationalism, region and regionalism, the issues are much less clear cut.

The concept of the nation state is so engrained in the consciousness of Western Europeans in the 20th century that it is tempting to assume that the two words are largely synonymous. As the contributions which follow indicate, however, it can be argued that, to a great extent, state formation has tended to precede nationhood which therefore becomes the result of a creative act, the fostering – if necessary the invention – of collective values and myths, the building of social cohesiveness (see Kedourie, 1960, Le Bras and Todd, 1981; Todd, 1990).

To complicate matters further, interpretations of 'nationalism' vary according to country, culture and language. In the Italian context, for example, 'nationalism' tends to mean the nationalism of the Italian state, while in Spain its use is much more flexible, so that no hackles are raised by references to the historic nations of Catalonia or the Basque Country, with their own collective values, their own cohesiveness. In the United Kingdom, the word 'nationalism' is just as likely to be applied to peripheral separatisms as to the state as a whole (Keating, 1988, p. 10).

Finally, 'region' and 'regionalism' pose their own problems as defining terms. The use of 'region' or 'regional' in a trans-national context, in the form of state-to-state relationships (e.g. Poland-Hungary-Czechoslovakia) is beyond the scope of this study, which will focus instead primarily on the 'region' as a unit smaller than the state which contains it (Waever, 1993). This may mean a territory given the status of a region for administrative purposes, a unit occupying an intermediate position between central and local government. It may also mean a territory having a claim to a cultural and political individuality of its own, marked out by ethnic, historical, linguistic features, moulded by shared myths and traditions. A region may, of course, display both characteristics.

The word 'regionalism' can denote the aspirations and activism of the concerned inhabitants of a region, and can usefully be applied to the pursuit of the specific interests of such a unit. Yet a note of caution is in order here too, since 'regionalism' can often be confused with regionalisation, which is perhaps better thought of as the pursuit of state-centred policies designed to impose 'top-down' remedies (especially economic ones) to regional problems and imbalances. To this extent, it may be argued that regionalism and regionalisation are mutually contradictory.

For our purposes, then, 'nation' and 'region' are to be seen as key elements in the centre-periphery model most frequently adopted to describe the evolution of countries and the formation of states. This centre-periphery model of state formation is generally used to explain the process by which influence is exerted by a central point on the surrounding periphery to create a reasonably cohesive state and society. The application of the model to the countries of Western Europe is persuasive, to the extent that cities such as London, Paris, Madrid, Lisbon, Rome, Vienna, and Dublin have clearly performed that function of central point in their respective territories in political, in economic, and sometimes in cultural terms, assimilating their peripheral regions. It is problematical in the case of Brussels, a capital sitting uneasily at the heart of a country which is riven by the stark distinction between two cultures and communities, Dutch and French. And for very different reasons, the model is quite inappropriate in the case of Germany, a polycentric, or polycephalic, state with no central point to act as a magnet or as a force for the diffusion of a dominant political or cultural message.

And if, in any case, the centre-periphery concept is taken less literally, and seen less in terms of spatial than of cultural relationships, then the process of diffusion by which the centre permeates throughout its extended territory becomes less convincing. Or at least peripheries may start to discover that diverse cultural, social and linguistic values are irreducible to a central uniformity, and that shared values and myths produce regional cohesiveness in its own right. The revival and rehabilitation of languages such as Breton, Welsh, Basque, and Catalan, restores a sense of identity to regions (nations?) long deprived of their very means of expression. The Gaudi-designed apartment blocks on Las Ramblas in Barcelona stand as a statement of regionalist faith. The entire question of relationships between nation and region, centre and periphery then becomes fluid and liable to redefinition according to place and circumstance, creating a dynamic challenge to the national status quo.

At first glance it may appear that, in the United Kingdom, the political climate has for years been less favourable to the cause of regionalism than in almost any of the other countries under discussion here. Yet, as Alan Butt Philip argues, a dismissive reaction to the issue of regionalism reveals more about the lack of regional awareness among the opinion formers of southern Britain than it does about the merits of the regionalist cause itself. The situation is further complicated by the tendency to focus on the nationalist demands of Scotland, Wales and Northern Ireland, at the expense of arguments in favour of the creation of regionalist institutions within England. This complexity in itself explains the reluctance of successive United Kingdom governments to become enmeshed in arguments about regionalism. The protracted debate on devolution during the 1970s, prompted in part by the growth of electoral support since the 1950s for 'minority' parties, all with some form of commitment to regional devolution, proved inconclusive. The failure of proposals for legislative devolution for Scotland and administrative devolution for Wales revealed the problems involved in ignoring the English dimension. The Conservative government of 1979 effectively removed the entire subject from the political agenda. It is clear, however, that the landslide election victory in 1997 of the new Labour party led by Tony Blair has brought the subject to the forefront of government thinking. The substantial majority of Scots in favour of plans for a Scottish Parliament, and even the narrow majority of the Welsh in favour of an elected Assembly for Wales indicate that the relationship between the constituent parts of the United Kingdom is in the process of significant change.

The apparent homogeneity of British society, underscored by nationally-organised mass media, can in fact be seen as a veneer concealing a bewildering diversity, in terms of origins, geography, language, religious observance, popular culture and the arts. In all, a multiple picture emerges of regions in the United Kingdom. There is a clear recognition of the distinctiveness of Scotland, Wales and Northern Ireland, while in England the absence of any sharply defined regional administration masks the existence of age-old local identities rooted in administrative and socio-cultural traditions which can be traced back for a thousand years.

Regional policy in the United Kingdom, with its origins in the 1930s, has, in contrast to the emphasis in other European countries on land use and infrastructure development, been pursued with the primary objective of maintaining or creating employment. It is hardly surprising, then, that in the years following 1979, Conservative ideology was opposed to such interventionist tactics, and regional policy was thus accorded a low priority. Only in Scotland, Wales and Northern Ireland, where regional development agencies have worked in concert with the regional departments of central government, has it been possible for policy to evolve beyond the scrutiny of central government, often to the envy of the English regions.

One effect of the dismantling of a regional tier of administration in the United Kingdom since the early 1980s was the need for local authorities to become fully aware of the opportunities offered to them by the regional policies of the European Community, providing a conduit for the submission of subsidy applications and requests for support from European Community structural funds, conversion programmes and the like. Ironically, the European Commission appeared from a certain perspective to understand regional needs and aspirations better than central government in Whitehall.

However, a political culture founded on the belief in firm central government as a prerequisite for the survival of the state and the reinforcement of British power throughout the world has left little room for the concrete, institutional expression of diversity. This, coupled with a predictable reluctance on the part of the Westminster-based élites to put their own authority at risk, has ensured that the regionalist debate has, until very recently, been largely conducted at the margins of British political life. At the end of the 1990s, however, and in spite of the apparent reluctance on the part of the English, as opposed to the Scots or even the Welsh, to embrace the idea of regional government, a regional tier of administration is being encouraged by central government. Already, politicians are faced with the choice of making a career in central government or at Scottish, Welsh or English regional level.

In Ireland, as Ullrich Kockel explains, the dominance of Dublin at a political and economic level has ensured that the centralisation inherited from the colonial past remains largely intact. With no significant role for the provinces, loyalties are felt at local or county, but not regional, level. Indeed, as far as European Union regional policy is concerned, Ireland has been treated as a single region since 1975, with the result that Dublin controls European funds allocated to the country as a whole. National development has therefore been seen as more important than regional interests, and such regional structures that exist are little more than executive agents of the centre. Problem areas are identified as – Dublin, with a declining city-centre economy and growing suburban population; the border with Northern Ireland, where deep-rooted enmity has prevented serious cross-border cooperation; and the Gaeltacht, or Irish-speaking areas, mainly along the western seaboard, where there is a conflict between economic modernisation and the wish to preserve and support the language. From the late 1980s onwards, there have been attempts to produce national plans with a regional element, so that Ireland can benefit from European Community Support Framework funding. However, the threat to that funding posed by the likely loss of Objective 1 status (the highest category of the Structural Funds budget) casts doubt on even the modest devolution which those plans contained. Future development of the European Union Committee of the Regions may help to loosen the control which Dublin exercises over the rest of the territory.

In France, the regional question is thrown into stark relief by the Napoleonic centralisation which has characterised the administrative, economic, and social structure of the country for the best part of two centuries. The argument that the French nation is a creation of the French state, or of a succession of regimes constituting the power of the state and extending back into the Ancien Régime, is a persuasive one. It is hard to escape the conclusion that the multiple divisions of life in the *Hexagone* – political, social, cultural, linguistic – were systematically suppressed by the mould of Republican uniformity. The concentration of political and economic power in Paris, and the consequent centralisation of decision-making, expressed through the prefectorial system of administration, held sway throughout the 19th and much of the 20th centuries. Only gradually, with the growth of industrialisation and the consequent unevenness of development, did the disadvantages of a rigidly hierarchical system of administration and centralised planning make themselves felt. The dearth of reliable information about economic progress in a vast, varied and relatively underpopulated country led also to an element of inertia, even complacency, which was not shaken until the years after the Second World War. This was the period of a growing realisation that the growth and prosperity of Paris had drained the French provinces of resources and manpower, leaving an enfeebled network of provincial towns and an impoverished agriculture. Policy thereafter was directed at a realignment of economic activity, so that the traditional, if oversimplified, demarcation line between a relatively prosperous and dynamic North, East and South-East, and an underdeveloped and stagnant West and South-West became less clearly defined. During the 1950s and 1960s too, the reassertion of regional identities began to be felt, predominantly in the more excentric provinces – regionalist movements in Brittany, Corsica, Occitania and, to a lesser extent, the Basque Country and Alsace, claimed attention with demands for varying degrees of cultural, political, and economic autonomy. The twin pressures of political and economic demands led to a gradual process of regional reform emanating from the centre, highlighting the paradox inherent in a siuation where central government takes decisions relating to matters of regional autonomy. Arguably the most significant reforms were set in train in the early 1980s when, for the first time, the principle of direct election to regional assemblies was conceded. Conclusive evidence of the effectiveness of these assemblies in improving the lot of their populations and in fostering the development of regional identity and allegiance has yet to be provided. Equally elusive is the sense that the strengthening of this intermediate tier of government has reinvigorated a political process increasingly seen as stale and unimaginative. There are nevertheless indications that, at the very least, local and regional élites are intent upon redefining their relationships with their traditional legislators and paymasters in Paris. New networks of contacts between regions sharing similar preoccupations and interests, both within and beyond the national boundaries, have opened up the prospect of shaking off, albeit gradually, centuries of Parisian centralisation in favour of wider transnational allegiances. It may be, of course, that the very centralisation against which the various French regionalisms have tested their strength provides the firm base which permits the contemplation of newer structures. It is, however, just as likely that the vision of interregional or cross-border cooperation, however tentative and exploratory, is seen as a welcome antidote to domestic political institutions and movements unable or unwilling to undertake a necessary process of self-renewal. It is certainly the case that, at the end of the 1990s, the principle role of the regions is expected to be that of economic regeneration and the creation of employment.

The process of state formation in Belgium, and the forging of national identity, has been problematical in the extreme. Indeed, the attempt, begun in 1830, to create a Belgian nation out of two culturally and linguistically distinct populations cannot be said to have met with unqualified success. Here, more than in any other of the countries presently under discussion, national unity is most fragile and most frequently under threat. Here, too, the language divide is most sharply focused. The frontier between Dutch-speaking and French-speaking communities, crossing Belgium like a geological faultline but acknowledged as a political reality only in the 1960s, has perpetually bedevilled attempts to bring into being a cohesive national identity. The role of the constitutional monarchy as a symbol of unity and amicable coexistence has the appearance of a veneer, always fragile, particularly at times of dynastic succession, despite the upsurge of popular and largely unexpected feeling generated by the death of Baudouin in 1993 which brought to an end the 40-year reign of 'the King of the Belgians'.

The uneasy relationship between the two communities was further highlighted by a reversal of economic fortunes which undermined the power of the minority francophone ruling élite and gave the Flemish community the economic weight to match its numerical ascendancy. With the old industries of the Walloon territory, coal, iron and steel, in decline, Brussels, Antwerp and the Flanders coastline thrived and grew. This shift in economic prosperity stimulated the demand, on the Flemish side, for an institutional recognition of their cultural identity and, on the part of the Walloons, the determination to defend an increasingly precarious status by setting in place the territorial limits of a francophone region. A complex series of constitutional revisions undertaken with increasing urgency from 1970 onwards has addressed these issues, culminating in the adoption in 1993 of legislation which, in its broad sweep if not in every detail, set Belgium firmly on the road towards federalism. In all this, the issue which has proved most intractable over many years is that raised by the ambivalent position of Brussels. A capital city located in Flanders, having authority over the entire country and governed by a francophone élite of long standing, Brussels is overlaid with patterns of allegiance, responsibility and identity of the utmost complexity. Little wonder then that successive Prime Ministers since the 1980s at least, have sought for Brussels the role of capital of a putative federalised Europe in order, as it were, to superimpose one pattern on another, lending legitimacy and conviction to a federal Belgian state which has yet to demonstrate its long-term stability.

Support for the process of European integration is equally strong in the Netherlands, partly because it has always been seen as a moderating influence on the power of Germany but also, as Bernard O'Sullivan and Dennis Linehan demonstrate, because national government is no longer felt to be best equipped to manage the economy, nor to redress regional imbalances. What emerges strongly is the consensual approach to the organisation of Dutch society, based on the long-established notion of co-governance, shared between municipal, provincial and national layers. From the 1970s onwards, traditional regional policy, based on redistribution and regional assistance from central government, was abandoned in favour of a policy which focused on key economic strengths and the promotion of public-private partnerships. Examples from the southern province of Limburg show how a traditional industrial region in decline was reinvented as a 'service-oriented Euro-Region', capitalising on its position between the growth areas of the Dutch Randstad, the Ruhrgebiet in Germany, and Brussels and Antwerp in Belgium. The creativity of regions able to develop their own economic priorities is thus seen as the key to prosperity in an integrated Europe.

The German perspective on federalism proposes lessons, too, on the process of European integration. Theo Stammen explains how the modern German state was formed in the ruins of the Third Reich and under the tutelage of the Allied Occupying Powers. The constitution of the *Bundesrepublik,* or Federal Republic, reflected both an awareness of earlier broadly federal traditions under first Bismarck and then the Weimar Republic, and also a determination to avoid the pitfalls of those early examples. In particular, the relative impotence of the regions under the Weimar Republic explained the weakness of resistance to Hitler's rise and permitted the all-too-easy dismantling of federal influence following his seizure of power. The fundamental importance of the *Bundesrepublik's* carefully balanced federal system as a defence against the repetition of past mistakes, against the concentration of too much power in one place, is underlined by the fact that the federal principle is unalterably enshrined in the constitution or Basic Law (*Grundgesetz*). The federal tradition extended also to the 'other' German state, the fledgling German Democratic Republic which, at least in its early years, adopted the Weimar pattern, only to abandon it once the goal of a monolithic socialist state was established. The federal state system in Germany offers a settled pattern of vertical power distribution through localities, *Länder* and the state, plus horizontal attributes of self-government at each level as appropriate. The allocation of responsibilities within this system is complex and carefully regulated, with clearly delineated powers for each level, as well as an element of concurrent legislation which permits shared decision-making in some areas. In addition, the redistributive nature of the tax regime ensures that those *Länder* which are the least well developed economically are effectively subsidized by the more prosperous. This measure of 'equalisation' of tax revenues is particularly significant at a time when the old north-south economic divide has been altered by the decline of traditional industries, and when the severely under-developed *Länder* of the former East Germany have been absorbed into the *Bundesrepublik*.

In political terms, too, federalism is seen as a crucial element in the building and reinforcement of the democratic process. From the origins of the *Bundesrepublik* a deliberate policy of education through democratisation was adopted, with citizens made familiar with the electoral process first at local level, then in the *Länder* and finally nationally, at the level of the *Bund*. In addition, with the allocation of responsibility for, in particular, education policy to the intermediate level of the *Länder*, a plurality of systems and programmes was encouraged. This plurality is in itself viewed as a positive value, working against any tendency to the despotic and the authoritarian.

There are two observations to be made about the German experience of federalism. The first relates to the current situation inside Germany itself and to the problems posed by the process of German unification. The euphoria which surrounded the initial stages of unification has evaporated, to be replaced by a disillusion rooted in an awareness of the costs of reintegrating the former East German *Länder* into the economy and society of the *Bundesrepublik*. While the difficulties are in no way attributable to federalism, it is nonetheless true that the size of the task has imposed enormous strain on a federal structure which for the first 40 years of its history was able to evolve in an atmosphere of economic growth and optimism.

The second observation is more positive and outward looking. For the process of European integration there is clearly a need for new models of political and institutional structures and organisation. Germany is the only country in the European Union, apart from Austria, with a developed and tested federal structure which, to an extent at least, subordinates the power of the state to that of its constituent parts. It can thus be seen as a pertinent example of possible future patterns of development for the European Union as a whole which, if it is to make further progress towards integration, may well have to consider ways of strengthening the diversity of regional identities within a wider institutional framework. Indeed, it is perhaps the German model which approximates best to the concept of 'subsidiarity' enshrined in the Maastricht Treaty.

Austria, too, has a federal constitution, although its provinces have less autonomy than their German counterparts. Josef Honauer shows that, in marked contrast to Germany, however, this is a monocephalic state, with the capital, Vienna, economically dominant and politically distinct from the rest of the more conservative *Länder*. In spite of Viennese dominance, the proximity of the Iron Curtain and the uneven development resulting from decisions taken by the Allied Occupying Forces after the war has led to the underdevelopment of East Austria. In terms of national and regional allegiance, Austria has existed for so long in the shadow of its larger neighbour that it is only in the recent years that a large majority of its citizens profess belief in its status as a fully-fledged nation. It is a matter of some concern that allegiance to and identification with the *Länder* is not reflected in meaningful democratic representation at that intermediate level.

In his discussion of Scandinavian regionalism, Lee Miles underlines the fact that the distinction between inter- and intra-state regionalism has often been blurred, as attempts to stress common Nordic approaches have been matched by an emphasis on differences between countries within the region. He argues that, while much attention has been given to the potential for Nordic (Baltic) cooperation, it is time to pay attention to those elements of intra-state regionalism taking place within those countries. Taking the case of Sweden, traditional concepts of a social democratic state with a highly developed public sector have been eroded in the last few years by pressures of liberalisation and deregulation. At the same time, the influence of the European Union has led regional authorities to work together, both within Sweden and with neighbouring regions in, for example, Denmark, in order to bid for European Union funding.

Italian regionalism is analysed by Anna Bull in terms of its administrative, political, and economic impact at different stages since Unification. Early debate about the merits of regional devolution revolved around the agreed need to harmonise regional differences and a wide range of cultures – the method finally adopted, in 1865, saw the introduction of a rigid prefectorial system which established representatives of national executive power at local level. The perception, in the aftermath of the Second World War, that the rise of fascism in Italy had been aided by the centralised nature of the state, led to a new involvement with the idea of decentralisation. Paradoxically, the re-emergence of federalism as a strand of political thought coincides with the attainment by Italy of a greater sense of cultural and linguistic homogeneity than at any previous period. Other factors, therefore, create a specifically Italian environment in which federalism and regional preoccupations can flourish. These include, above all, the corrupt nature of the relationship between the political and business worlds, organised crime, and the general deterioration of political institutions, all serving to induce cynicism on the part of the electorate and a desire for systematic institutional change. The Northern League's proposal for a federal state made up of three macro-regions (North, Centre, South) won an unexpected level of electoral support at the beginning of the 1990s and, with the disarray to which the Socialist and Christian Democratic parties have been condemned by the recent corruption scandals, appears to have set part of the political agenda for the immediate future.

Throughout much of the 20th century, attempts have been made to address the problem of the under-development of southern Italy. The two-way process of state subsidies in return for electoral support was a constant feature of the political scene. Attempts to help first agriculture and then industry, through the centralised agency of the *Cassa per il Mezzogiorno* had limited success, with problems blamed on excessive centralisation, yet subsequent attempts to devolve responsibility became mired in the ever-present corruption and clientelism. Accordingly, the concept of regional development in itself has been called into question by Northerners resentful of the need to subsidise the South, and this resentment can be seen as responsible in part for the federalist proposals of the Northern League. Legislation enacted in 1990 to regulate relations between central, regional, and local government appears to assign superior powers to the regions in planning terms, and yet federalist demands persist. Popular support for institutional change, embracing regional reform, was indicated by the results of a number of referenda held in 1993, although the speed at which political developments have occurred during this period makes the future highly unpredictable. By the elections of 1996, the stance of the Northern League was frankly secessionist, and all the main political parties appeared to be supporters of federalism, even though each chose to define the term differently. Laws enacted during 1997 both enhanced the responsibilities of regions within the unitary state, introducing the concept of subsidiarity, and decreed that local or regional autonomy should vary according to the competence of the authorities concerned.

European integration receives broad support from the Italian population as a whole – indeed, political groupings such as the Northern League see the European dimension as an ideal context for the renewal of contact between Northern Italy and its trans-alpine neighbours. At an economic level, trans-national and interregional collaborative agreements have met with wide support. The conclusion is, however, unavoidable, that Italian enthusiasm for many or all aspects of European integration may have less to do with trans-national solidarity and idealism than with a widespread distrust of and frustration with central government. It is the perceived inadequacies of a particular – Italian – form of the nation state that explain the appeal of an as yet ill-defined federal Europe.

One of the main factors inhibiting regional development in Greece has been the extreme centralisation of the Greek state, while regional disparities remain very pronounced. Dimitrios Christopoulos argues that there is a lack of strategic vision in regional planning, although eligibility for European Union regional funding is an incentive to modernisation. Since the early 1980s, centralism has been seen as a cause of the clientelism which bedevils the relationship between the capital, Athens, and regional urban and rural centres in what is a largely agrarian country. It seems clear, however, that the political élite at national level is moving towards a gradual rationalisation of adiministration at least, not so much because of pressure from the grass roots as from the need to conform to European directives and demonstrate probity. Laws enacted in 1997 intending to consolidate local authority structures may shortly be followed by the creation of an elected regional tier. Greece is thus seen as a unitary state obliged to contemplate a process of regionalisation in response to external (European Union) requirements. There is no indication that progress in this direction can be attributed to demands for local democratic accountabilty.

Jesús del Río Luelmo and Allan Williams offer two sharply contrasting images of the regional question in the Iberian peninsular. They trace the origins of the contrast to the difference in the process of state formation in Spain and Portugal respectively. In more than 500 years, the supremacy of the Portuguese state has remained virtually unchallenged by any of its regions. This is attributable to the fact that the state, formed by military conquest over the Moors from the 14th century, preceded the nation, which duly found its identity, and created its myths, in the challenge and opportunities of its Atlantic seaboard. Discovery, conquest and the founding of colonies brought unity and a sense of shared purpose. Furthermore, the ubiquity of the Portuguese language throughout the territory has cemented that unity, underlining once more the central importance of a distinctive language as an indicator of cultural limits and thus of national identity. Other factors also impinge on Portuguese nationhood to various degrees – an urban network in which the influence of Lisbon is preponderant; the pervasive influence of political and economic élites from the urban bourgeoisie and their gravitation towards the capital; the absence of significant minorities – these are all factors which go some way to explaining the homogeneity of the Portuguese state. Clearly, the half century of dictatorship under Salazar confirmed and reinforced this seamless national identity, with its high degree of centralised, corporatist control and minimal financial power at local government level. Even after the transition to democracy in the mid-1970s, however, tentative plans for a regional administration stayed very much in the shadow of a reinvigorated local, municipal government, fostering the view that the size and uniformity of the country as a whole does not justify an additional, intermediate tier. The sole exception to the state's unquestioned supremacy has been evident periodically in the Atlantic islands, but the granting of directly-elected assemblies to the Azores and Madeira, with the power to raise local taxes, in the post-Salazar era has largely defused the impact of their regionalist movements, despite the islands' reliance on Lisbon for budgetary supplementation.

In Spain, however, the picture is very different. Successive waves of separatist activity since the early 19th century form a backcloth for modern regionalist aspirations. State formation in Spain has a long history, and is rooted in both military conquest and alliances through royal marriages. This latter feature ensured, in contrast to the situation in Portugal, the persistence of a variety of regional identities outside centralised Castille. Ancient rights, notably in the Basque country and Catalonia, coupled with uneven economic development from the very beginnings of industrialisation, led to regionalist pressures from the late 19th century, and to the restoration of a measure of autonomy in the three historic minority nations, Catalonia, the Basque Country and Galicia, in the early decades of the 20th. Subsequently, the Franco regime eradicated all traces of regionalism, but failed to expunge the desire for it, so that, on Franco's death, demands for regional recognition and reform were loud and persistent. A number of factors provide an explanation for the strength of these demands. The most far-reaching and fundamental of these is cultural difference, exemplified by the language factor, which differentiates and identifies Catalans, Basques, and Galicians. Castillian Spanish is spoken by no more than three quarters of the population of Spain as a whole, with the result that the three minorities and, in particular, large cities such as Barcelona or Bilbao, exert a centrifugal force, encouraging the growth of regionalisms and thus of a society with a number of different focal points.

Cultural difference was amplified in the mid-20th century by starkly uneven economic development. With the country's political centre, Madrid, relatively undynamic in economic terms, two of the three industrially active regions – the Basque Country and Catalonia – had formed the powerhouse of early industrial development. Much of western, central and southern Spain remained undeveloped and poor. The negative effects of this pattern of development were multiple: imbalances were accentuated by migration from poor to prosperous regions; the poorest regions posed problems when the time came to create some form of regional equity, and the successful regions such as Catalonia objected to subsidising their less successful neighbours. Paradoxically, Franco's regime itself offered succour to the regionalist movements, to the extent that its heavy-handed suppression of regional activity encouraged a spirit of anti-government solidarity between regionalist movements. With the return to democracy in the post-Franco era, a measure of autonomy was seen as an essential element of constitutional reform, but its implementation posed a number of problems, not least in that some regions were obviously self-defining and eager to assume an autonomous role, while elsewhere it was by no means clear what, in precise territorial terms, constituted a region. The principles of state unity, regional autonomy, and interregional solidarity enshrined in the 1978 Constitution represented an attempt to deal pragmatically with widely divergent levels of concern. Subsequent legislation, and regional elections, have gone some way towards producing an equitable settlement which, despite its lacunae and drawbacks, has brought about, in less than 20 years, an astonishing degree of regional reform. Reservations remain about the strength of regional identity in some parts of the country, although the very process of reform has had the effect of raising regional consciousness in, for example, La Rioja and Cantabria.

Notions of allegiance and identity are difficult to define and assess, but the attachment of both the Spanish and Portuguese people to the idea of a European identity seems consistently high. However, it is clear that, while this attachment is actively encouraged by, and tends to be channelled through, existing regionalist movements in Spain, Portugal's unity as a nation-state is likely to remain unchallenged.

Yet the categories of allegiance, the sense of communal belonging, which arouse the enthusiasm of peoples are not immutable. Nor are they susceptible to easy definition, their origins, as often as not, lost in the mists of time. A consistent theme to emerge from the various essays in this volume is that of contrast between, on the one hand, state formation, the process by which communities, provinces, regions can be fashioned into cohesive entities, able to be governed and administered and, on the other, the search for regional identity, the assertion of diversity. Military conquest, political and dynastic treaties and alliances, are the tools which permit or make necessary the process of state creation. Centralisation of power, often perceived from the margins as internal colonialism, is the factor which provokes regional dissent and the desire for a distinctive voice.

A further theme which recurs in the discussion of regional identities and regional politics is that of a movement away from standard centre-periphery relationships, with their self-perpetuating, oscillating tensions between state and region. It is a movement which can perhaps best be seen as the attempt to define a new regionalism, putting to one side the dictates of national allegiance, seeking to create new, trans-border regions, forming political, cultural, economic networks.

Such networks clearly find favour, particularly in the more obviously peripheral regions of the states under consideration here. The Mediterranean arc which links

Catalonia with Piedmont and Lombardy via the southern French territory of Occitania; the Atlantic arc which brings together Wales, Brittany, Aquitaine, Galicia and establishes common cause between the maritime towns of Oporto, Bilbao, Nantes and Milford Haven; Anglo-Franco-Flemish declarations of common economic and social interest – all these ventures bear witness (for all their occasionally implausible connections) to the desire to create new ways of belonging. Unlike the nation or the region, however, these newly-conceived trans-border regional networks suffer from more than one serious lack – their very newness deprives them of a clear identity and makes it hard for them to arouse in their populations any sense of belonging or allegiance. Any attempt to forge new identities, new patterns of allegiance, must take these needs into account (Smith, 1992, p. 73). Furthermore, they lack the legitimacy that comes with democratic accountability; their leading figures are, at best, genuinely representative only of their own constituencies.

What are the factors that govern the sense of belonging? How is the identity, and hence the allegiance, of an individual, or of a community, determined? There are perhaps two principal sets of criteria to be taken into account. First, identity lies in memory, the individual memory of a past life and experience, the collective memory of tradition, shared belief, myths, symbols, expressed in culture, ritual, and language. The classic regions and nations which feature in this book patently conform, in varying degrees, to many if not all of these criteria. The second concerns the definition of identity by reference to an external reality – the individual, the collectivity, has an identity defined in part by its limits, by what it is not, by the Other. Here again, region and nation fit the pattern, the nation consolidating its form by contrast or conflict with its neighbour(s), the region by asserting itself in relation to its centralising state.

Both definitions call into question the identity of new formations. As they depend on the accumulation of remembered experience, they offer no obvious support for those structures which reach tentatively into new areas and attempt to create new patterns. Trans-border regions have no easily reconstituted history, no common language or shared mythology. Even more amorphous is the overarching concept of a European identity, founded on a Babel-like confusion of signs and traditions. The only collective memories held in common by the nations and regions of Europe are memories of strife and bloodshed. Little wonder then that, in large measure, the people of Europe hesitate to embrace a European identity, when there is such an enormous gap between the fine, utopian ambition of supra-national allegiance and the mundane hotchpotch of Regional Development Funds, European Social Funds, Orientation and Farming Support Funds, and the like.

There are dangers here. The exhortation to participate in a European future risks the creation of an identity vacuum, or at least an insecurity which, particularly in times of economic recession, reduces the sense of identity to the level of tribalism, an atavistic instinct which asserts itself against others in the crudest way. The growing support in recent years for the Extreme Right throughout Western Europe, and particularly in France, Belgium and Germany, stimulated in part, at least, by the perceived threat of mass migrations from the South and East, is evidence enough of the trend.

There is, though, a more positive aspect. The same mobility which brings about the clash of cultures and living standards also ensures that, increasingly, individuals have an interest or a stake in a variety of communities, large and small. Overlapping circles of allegiance, networks of belonging, are perhaps the most appropriate way to describe the formation of identity in late 20th-century Europe. In true post-modern fashion, everyone can be many things simultaneously, 'baker, railway enthusiast, mother, conservative, from Hamburg' (Waever, 1993, p. 207). Of course that has, in many respects, always been the case. Throughout much of the 19th and 20th centuries, however, it is the single category of nationality that has held sway, and that continues to exert both influence and appeal over the people of Europe, waxing and waning according to time, place and the vicissitudes of the economic cycle like the gravitational pull of some moon in its elliptical orbit, stronger then weaker as the seasons pass.

One of the consequences of globalisation, as the end of the 20th century approaches, is the realisation that the shapes and contours of the world we inhabit, the boundaries which define our culture and sense of belonging, are fluid rather than fixed, constructed rather than given. These boundaries remain the means by which we classify and impose meaning on our world and define our own identity, collectively and as individuals, yet we may, in acknowledging their fluidity, open up new modes of belonging without threatening to supplant the old. It is tempting to conclude therefore that, however strong the gravitational pull of nationality may be, many smaller moons are also visible, with orbits which overlap. The resulting patterns and relationships may then lead to multiple allegiances and a form of belonging which enables a Scot from Glasgow to be European, just like a Basque from Bilbao, or a Breton from Brest.

References:

Henri, comte de Paris, (1996), *Les Rois de France et le sacré*, Monaco, Editions du Rocher.

Ignatieff, M. (1993), *Blood and Belonging: Journeys into the New Nationalism*, London, BBC and Chatto and Windus.

Keating, M. (1988), *State and Regional Nationalism: Territorial Politics and the European State*, London, Harvester Wheatsheaf.

Kedourie, E. (1960), *Nationalism*, London, Hutchinson.

Le Bras, H. and E. Todd (1981), *L'Invention de la France: atlas anthropologique et politique*, Paris, Librairie Générale Française.

Todd, E. (1990), *L'Invention de l'Europe*, Paris, Seuil.

Rokkan, S. and D. W. Urwin (eds) (1982), *The Politics of Territorial Identity: Studies in European Regionalism*, London, Sage.

Smith, A. D. (1992), 'National identity and the idea of European unity', *International Affairs*, 68 (1), 55-76 (p. 73).

Waever, O. (1993), 'Europe since 1945: crisis to renewal', in van der Dussen, J. and K. Wilson (eds), *The History of the Idea of Europe*, Milton Keynes, The Open University.

Regionalism in the United Kingdom

Alan Butt Philip

Introduction

Many observers of British society and much of public opinion would deny that regionalism is an issue in the United Kingdom in the 1990s. Such a standpoint reflects a lack of regional consciousness in much of southern Britain and a lack of knowledge of, or even sympathy with, other parts of the British state. Electors often need to be reminded that the United Kingdom is comprised of four nations – England, Scotland, Wales and part of Ireland, as well as the Channel Isles and the Isle of Man; or that indigenous languages other than English are spoken in Britain – particularly Welsh (spoken by around 500,000 people) and Scots Gaelic (spoken by about 80,000 people). For the purpose of this chapter, the nationalist demands of Scotland, Wales and Northern Ireland will be considered alongside the less obvious regional demands of England itself, even though the arguments for separate recognition and demands of different national identities are of a much larger scale and ideological order than those advanced for regionalism all round. In any case, development within the British constitutional system of stronger political institutions for the different nations of Britain is increasingly seen as contingent upon the acceptance of a set of regional institutions for England. Despite a public rhetoric which often denies and denigrates regionalism (at the level of national government and the tabloid press), there are several indications that the regional issue will not disappear and that it is having to be addressed, albeit piecemeal, by central government. As the crisis of legitimacy facing Britain's political institutions and élites has deepened, so the regional issue broadly defined has returned in the 1990s to occupy a central place on the political agenda, just as it did for a time in the 1970s. As we approach the millennium it begins to look as if tangible progress towards devolution and regionalism in the United Kingdom will at last be made.

Origins and History

The origins of the United Kingdom's reluctant entanglement with regionalism range from the nationalist claims of Scotland and Wales which have had to be accommodated by the British political system since the 1880s; the political settlement of the Irish question in 1921 which led to the creation of a largely self-governing province of Northern Ireland; and the practical administrative and occasionally decentralist demands of the English government and political class. The regional dimension in

British politics has ebbed and flowed, but it has never died away completely. The spread of political allegiances in the last quarter of the 20th century suggests that it is likely to be a continuing, yet sporadic, flashpoint in the British political system.

One of the spin-offs of the paralysis of the United Kingdom regime caused by the campaign for Irish home rule in the 19th century was the seriousness with which claims for Scottish and Welsh home rule were taken. The Liberal platform of the late 1890s contained promises of 'home rule all round', a package of devolution measures affecting the whole of the British Isles, even if a precise role for the English regions was not defined. Progress on the Irish question was frustrated by the House of Lords and by the outbreak of the First World War. By 1918 it was clear that a clean break would have to be made between Ireland and the rest of the United Kingdom, an out-turn which choked off the broader British debate on devolution. The nationalist cause in Scotland and Wales became channelled into political parties dedicated to self-government formed in the 1920s and 1930s. A petition (the Scottish Covenant) demanding a Scottish parliament was signed by two million people in the late 1940s, and a similar but much less widely supported petition to parliament was organised in Wales between 1950 and 1955 (Hanham, 1969 and Butt Philip, 1975). No one at this stage was marching for the cause of English regionalism, nor have they done so since.

The current cycle of devolutionist pressures in the United Kingdom can be traced back to the mid-1960s when a combination of regional economic decline, disillusion with the Labour party and anxiety (in Wales) about cultural change fed a strong political revival of Scottish and Welsh nationalism. The House of Commons was forced to spend much of its time between 1974 and 1979 on devolution legislation which, however, came to nought as a result of two referenda held in Scotland and Wales in March 1979. Matters in the province of Northern Ireland were also on the move, but in an unrelated way. The inability of the Protestant dominated Unionist Government in the province to deal adequately with grievances from the minority Catholic population caused the Conservative Government at Westminster under Edward Heath to abolish the provincial government at Stormont in 1972 in favour of direct rule from London. Over 20 years later this remains the position, despite many abortive attempts to re-establish an elected Northern Irish Government based on power-sharing or with entrenched minority rights.

The agitation for greater recognition of Scottish and Welsh devolution has had other political consequences, such as the transfer of new powers and resources to the Scottish Office and the newly-established Welsh Office (1964), the passage of the Welsh Language Act (1967) giving equal validity to Welsh in public administration in Wales, the mobilisation of MPs more frequently into regional groupings to lobby their own party leaderships, and the setting up of a Royal Commission on the Constitution under Lord Kilbrandon, which reported in 1973. Regional policy also attracted greater attention in the 1960s and 1970s before being cut back by Conservative administrations after May 1979. Local government and administration has been a continuing object of attention by successive Tory prime ministers. The formation of the Greater London Council (GLC) and the six metropolitan county councils in England in 1963 was followed by a wholesale reform of local government in England, Scotland and Wales in

1973-4, abolition of the GLC and the metropolitan counties in the mid-1980s, progressive privatisation of service provision previously undertaken by local councils, and further reviews of local government structure in 1992-4 in order to establish unitary local authorities as widely as possible. Such turmoil is symptomatic of an early phase of institutional reforming zeal, reflecting confidence in the ability of structural changes to improve quality of service, which gave way in the 1980s to a zeal to cut down bureaucracy and to contain burgeoning local government spending. Regional strategic needs and the desirability of creating or building upon existing regional identities have been very low on central government's list of political priorities. Yet the regional dimension has been hard to extinguish, even in England, and shows signs of revival as the millennium approaches, despite the increased centralisation of power in Whitehall and Westminster and the impact of improved physical and media communications upon the attitudes, place of residence and lifestyle of the citizens of Britain.

The Devolution Debate

In 1951 the Conservative and Labour parties captured between themselves almost 97 per cent of the total votes cast at the October general election that year. By February 1974 a quarter of the votes cast at the United Kingdom general election were cast for parties other than the two largest parties, most of them committed to some form of regional devolution or even more radical reorganisation of the British state. That position has remained broadly unchanged at all subsequent general elections. The vagaries of the British electoral system have meant that these votes have often not been reflected in the distribution of seats in the House of Commons. Yet between 1974 and 1979 the votes of Liberal, Nationalist and Ulster MPs played critical roles in ensuring the survival of Labour Governments in two successive parliaments, and there have been signs since the general election of 1992 that the 'minority parties' are once again becoming critical determinants in the passage through the Commons of significant parts of the Conservative government's programme. With such influence comes the power to re-set the political agenda to include issues related to devolution and regional policy (Rose, 1992 and Bogdanor, 1979).

The long drawn out attempt to establish assemblies in Scotland and Wales in the 1970s dominated the legislative timetable of the House of Commons and proved extremely divisive for Labour, whose northern English MPs were extremely suspicious of the advantages that would accrue to Scotland and Wales once devolution was in place. The proposals for a scheme of legislative devolution for Scotland and administrative devolution for Wales were eventually submitted to national referendums in each country. The Welsh scheme was roundly rejected by voters by a margin of almost four to one, while the Scottish scheme was approved by a small majority but failed to attract the support of 40 per cent of the electorate as required by the Westminster parliament. The whole exercise proved extremely debilitating for the political class as a whole, for no tangible result other than the ultimate demise of the Callaghan government. Yet the arguments had revealed some serious difficulties in trying to set up a scheme of devolution which omitted any consideration of the English dimension. Scottish and Welsh MPs at Westminster would have been able to vote on issues only of concern to England, because devolved assemblies in Scotland and Wales would be responsible exclusively for several policy fields. Northern and south western MPs were alarmed at the autonomy offered to Scotland and Wales but denied to their regions. The Welsh electorate proved very suspicious of anything seen as a sop to Welsh nationalism, and would have been much reassured if the devolution scheme had been part of a package in which the English regions were included.

The incoming Thatcher administration in 1979 was not only ideologically opposed to devolution but had compelling practical reasons for abandoning the whole project. The price of so doing was a progressive alienation of Scottish public opinion during the 1980s. Labour meanwhile was left to sort out its internal differences. No mention of a Welsh assembly, let alone a scheme of regional government for England, was made in its 1983 election manifesto, but by 1987 pledges of action on both had returned. The experience of the 1970s persuaded the Liberals that a federal solution was the only way to avoid the many anomalies that would be introduced by a scheme of piecemeal devolution. Unfortunately, some of their allies in the newly formed Social Democratic Party (SDP), such as Dr David Owen, were contemptuous of federalism – at British or European levels – and the Alliance leadership had to be content to reaffirm the primacy of establishing Scottish and Welsh parliaments as part of a phased introduction of an overall devolutionist structure.

The threat of rising Scottish and Welsh nationalism appeared to have been contained by the Conservatives for much of the 1980s, with Scottish National Party (SNP) representation at Westminster reduced from a high point of 11 MPs in 1974-79 to 3 MPs at the 1987 general election. Scotland continued to be the most promising lever on the British constitution upon which the hopes of all those who argued for home rule, regional devolution and a wider agenda of constitutional reform rested. In the United Kingdom as a whole, the cross-party Charter 88 movement was set up in 1988 to argue for a whole set of inter-locking constitutional reforms, including greater regional autonomy, and attracted 40,000 subscribers and some centre-left quality newspaper support. In Scotland itself, a broad-based coalition of forces, including the Labour and Liberal Democrat parties, the Church of Scotland and business leaders established a Scottish Convention which met to work out an agreed scheme for a new Scottish parliament. Despite the weakness of the Conservatives north of the border (with 9 MPs at Westminster out of 72), the government of John Major, who succeeded Mrs Thatcher as Prime Minister in November 1990, was able successfully to tar all the non-Conservative supporters with an anti-unionist brush at the April 1992 general election. Against all expectations the Conservative vote increased slightly to 26 per cent of those voting in Scotland; 10 Tory MPs were returned to Westminster and the status quo was defended and maintained. Further progress on devolution in Great Britain rested upon the Conservatives losing their majority in the House of Commons and this required a change of heart in England.

With hindsight the work of the Scottish Constitutional Convention during the 1987-92 parliament was to prove crucial to the re-emergence of the broad regionalist agenda after the landslide victory of Tony Blair and his modernised Labour party in the 1997 elections. Agreements made across interested political parties in these earlier years (but not including the Scottish National Party or the Scottish Conservatives, of their own volition) enabled a very speedy legislative progress to be made on Scottish devolution by the new Blair government, commanded widespread support among Scottish public opinion, and set the pattern for cross-party cooperation between Labour and the Liberal Democrats on the whole range of constitutional issues which formed a centrepiece of new Labour's manifesto. This cooperation was crystallised by the pact between representatives of the two political parties, Robin Cook and Robert Maclellan, five months before the 1997 election, under which they agreed to support a very ambitious package of constitutional reforms over the course of two parliamentary terms. This was followed up by an invitation to the Liberal Democrats after the election, which was accepted, to participate in the membership of a special cabinet committee overseeing the progress of this package through parliament.

The results of the general election of 1997 completely transformed the prospects of devolution and regionalism in the United Kingdom. The main winners, Labour, and to a lesser extent the Liberal Democrats, were both strongly committed to action in this field while the outgoing Conservative government lost all the seats it was defending in Scotland and Wales, as well as in Cornwall, and major cities such as Leeds, Sheffield, Birmingham and Bristol. The Blair government taking office with a massive parliamentary majority, a demoralised Conservative opposition and clear manifesto commitments to launch devolution, decided to waste no time. Proposals to set up a Scottish Parliament with legislative and tax-raising powers, and a Welsh Assembly without such powers, were rushed through parliament in a matter of weeks. These were then submitted for confirmation by referendum of all electors in each country in the autumn of 1997. The proposals for Scotland were approved by nearly three-quarters of voters in a high turnout election, although the proposed tax-raising powers were less well supported, despite the Prime Minister undertaking that Labour would not be using them initially. The Welsh referendum was much less clear-cut in its result, with barely half the electors casting their ballots and approval for the package obtained by a margin of just over 7,000 in a poll of over one million. Indeed in Wales there had been far less of a consensus-building debate about the structure and role of any devolved body, and it proved difficult to generate enthusiasm for such a relatively weak forum. Elections to the devolved bodies will take place in May 1999 using a German-style system of proportional representation. Thereafter these bodies will oversee and have some responsibility for most of the functions exercised by the Scottish and Welsh Offices, even though the Secretaries of State are also to be retained at United Kingdom government level, with seats in the Cabinet. The same position will face the Secretary of State for Northern Ireland when, following referenda conducted throughout the island of Ireland in May 1998, a devolved Northern Ireland government drawn from an elected assembly takes over many of the functions that Whitehall and the Northern Ireland Office have exercised since the demise of the power-sharing administration of 1974.

The phased introduction of regional institutions in England to complement developing devolution elsewhere in the United Kingdom was clearly spelled out in

Labour's 1997 election manifesto. Special treatment was to be given to London in order to restore an elected strategic planning authority to the capital, abolished by the Thatcher administration in 1986. The final scheme, for an elected Mayor of London supported by an elected authority (with responsibilities for economic regeneration, transport, planning, police and environmental protection) was supported by over 70 per cent of London voters in a referendum in May 1998, although only 30 per cent of the eligible electors cast their votes. The rest of the English regions are however destined to follow a much more cautious path towards devolution. A White Paper issued in January 1998 by the Deputy Prime Minister, John Prescott, in his role as Secretary of State for Environment, Transport, and the Regions confirmed that a step-by-step approach to English regionalism was being planned by the new government, starting with the creation of regional development agencies serving the existing eight administrative regions, excluding London. These will be complemented by the creation of regional chambers which will represent local political, economic and commercial interests and which are set to coordinate transport, planning, economic development, land use planning and bids for European Union funding. These may be the forerunners of elected regional assemblies which will only be created if there emerges a strong public demand for them, taking each region separately, and it is backed by a referendum in support. Local government representatives in many parts of England started to set up such regional chambers in anticipation of the legislation after the 1997 general election and have been encouraged to do so by the minister of state responsible, Richard Caborn, who has emerged as an enthusiast for regionalism all round.

A further twist, adding to the new regional configuration of England, came with the new government's decision to change the system for electing British members of the European Parliament. With effect from May 1999, MEPs will for the first time be elected to serve large multi-member regional constituencies (once again the English administrative regions, London, Scotland and Wales). So regional interests stand to be represented directly in the European Parliament in a structured way.

Regional Policy and Administrative Regionalism

Regional policy, defined as the attempt by governments to influence by policy measures the course of economic development in the regions, has a pedigree in the United Kingdom that goes back 60 years. The principal objective of regional policy has been to maintain or to create employment, whereas in other European Community states other objectives have formed part of regional policy, such as land-use and infrastructure planning. Macro-economic policy considerations arising from the danger of overheating the most successful regional economies causing cost-push inflation across the whole national economy have also justified government attempts to rebalance economic growth patterns between regions all over Western Europe.

The first attempts at United Kingdom regional policy were made in the Special Areas Act of 1934-7 in response to the collapse of employment in the slump years in regions such as South Wales, parts of Durham, Tyneside, West Cumberland and Scotland. This piecemeal approach was developed into a more generalised regional policy following an influential wartime review, the Barlow report. A more or less integrated regional, industrial and employment policy was maintained from the 1940s until the end of the 1970s and the arrival of Mrs Thatcher as Prime Minister (Smith, 1989).

The methods adopted by central government policy-makers have been used continuously but are often varied in terms of policy mix. The most favoured method has been the provision of capital grants to industry, in the private sector as well as in the public sector, linked to regional development and the creation of jobs. This approach has run into several difficulties. It has usually been addressed to manufacturing industry, when the main sources of job creation were in the service sector. It is increasingly constrained by European Community rules (see below), and its efficacy has been strongly challenged by neo-liberal economists. However a major study for the Department of Trade and Industry in 1986 found that much of United Kingdom regional policy has been successful in creating nearly 600,000 long term jobs (Moores, Rhodes and Tyler, 1986).

Another instrument that has been used to boost regional development and employment has been the provision of large public funds for infrastructure in the regions – for motorways and roads, rail electrification, telecommunications, power generation and water supply developments, new towns and, to some extent, new universities. Central government departments have been dispersed in whole or in part to regional locations away from London – the Crown Agents were sent to East Kilbride (near Glasgow), the main social security activity (now the Benefits Agency) went to Newcastle, the Department of Health to Leeds, and the Manpower Services Commission (now part of the Employment and Training Agency) to Sheffield. Economic development institutions have been set up such as the English Estates Corporation (now privatised) which built and managed industrial estates, new town and later urban development corporations, regional development agencies in Scotland, Wales and Northern Ireland, and the short-lived National Enterprise Board (1974-9) (Butt Philip, 1978). Government grants to subsidise current employment costs in regional undertakings played a major part (notably the Regional Employment Premium and the Temporary Employment Subsidy schemes) in the policy of the 1970s, and these continued in the 1990s in the public transport field with annual subsidies to the regional operational costs of British Rail and Caledonian MacBrayne's ferry services to the Scottish islands.

The ideological thrust of Mrs Thatcher's governments in the 1980s was clearly in conflict with such interventionist subsidies. Market forces could have been allowed free rein and regional dispersal of firms and employment achieved in the long term as companies moved away from high cost, scarce labour regions to lower cost, rich in labour regions. There were bound to be short term costs in terms of socio-political damage, leapfrogging wage and housing price spirals, and inadequate infrastructures in the expanding areas. Fortunately the theory was never fully tested as the rigidities in the labour and housing markets soon became clear; the famous 'get on your bike' advice given to the unemployed by one senior Tory Government minister, Norman Tebbit, fell foul of the realities of the absence of affordable and available housing in those areas where employment opportunities were greatest. The Thatcher government therefore never completely abandoned regional policy. They proved fearful of applying their economic philosophy fully in this area, and they were trapped by continuing European Community funds for regional development being tied to co-finance by United Kingdom central or local government. After major reviews of regional policy in 1983 and 1988, the 'social' justification for regional policy was accepted, but the budget for regional grants continued to be cut in real terms, and the areas eligible to receive such grants greatly reduced in size. Assisted area status was still much sought after in the 1990s, as the key to unlock funds from Brussels, if nothing else, and by 1993 the United Kingdom government was adding areas like Portsmouth, Thanet and the East End of London to its list having previously removed areas such as mid-Wales, North Devon and North East Scotland (Tighe, 1993 and Wintour, 1993).

Conservative governments may have downgraded regional policy as such since 1979 but they have accepted the case for limited public sector intervention in special areas, where possible linked with private sector finance. This has led to the formation of new designations such as tax-relieved 'enterprise zones', and new appointed bodies directly funded by central government, such as the London Docklands Development Corporation and a host of urban development corporations, operating for example in Bristol, the Black Country and Merseyside. Inner city areas have had to compete for central government funds in the 30 'City Challenge' schemes. In 1993 some unification of these separate urban initiatives was announced in the form of new 'City Pride' teams linking local businessmen with civil servants to oversee urban policy implementation (Willman and Burt, 1993). This represents another sign that regional policy is being overtaken by urban policy, and specifically inner city concerns. The bare economic indicators, especially the figures on regional unemployment rates (see Table 1), help to explain why this is so. In parallel with the onset of the 1989-93 recession there has been clear evidence of convergence in the economic circumstances of the various United Kingdom regions, although a similar trend predates this. In the 22 years since 1971 all regions saw a rise in unemployment, but whereas the most affected region (Northern Ireland) saw its unemployment rate double from 7 to 14 per cent, the least affected region (the South East) saw its unemployment rate rise almost fivefold from 2 to 10 per cent. The regional differential had thus narrowed over two decades, and this was also true for the movement of regional GDP, although the greatest change occurred as a result of the most recent recession, whose long term impact cannot yet be assessed.

To an important extent, the condition of regional economic policy remains stronger in Scotland, Wales and Northern Ireland, where specific initiatives have been made and separate political institutions exist to support and implement them. One reading of the economic convergence noted above is that the separate regional policies of those named areas have indeed proved their worth. In 1966, as a result of regional economic and political pressure, a new Highlands and Islands Development Board was set up to stimulate employment and economic development. This was followed in 1975 by the new Scottish, Welsh and Northern Ireland Development Agencies. They offered an integrated, interventionist, hands-on approach to the resolution of long standing regional economic difficulties and appear to have had much success. (The situation in

Northern Ireland has proved move difficult than in Scotland and Wales, possibly as a result of the continuing communal conflict there, although the Local Economic Development Unit (LEDU) which assists small firms in Northern Ireland has been very successful.) Early Conservative attempts to abolish them in 1979-80 foundered on intense local opposition, including from their own supporters. In the 1990s, the Scottish agencies have been broken up into smaller area-based units and they have been given employment training as well as economic development roles. They may however have lost much of their value as developers and implementers of regional industrial strategies, due to their smaller scale and local focus.

Undoubtedly a major element in the success of these development agencies has been their close sponsorship and relations with the so-called 'regional' departments of central government – the Scottish Office, the Welsh Office and the Northern Ireland Office, the latter set up shortly after direct rule was re-imposed in 1972. This administrative arrangement has transferred most of the domestic responsibilities of central government in these named areas to the relevant 'regional' department (e.g. agriculture, health, transport, education, environment and employment). The regional departments have enjoyed considerable autonomy, as well as the responsibility for distributing their sizeable block grants for public expenditure which Whitehall has handed down. This has enabled significant differences in the evolution of education policy or the levels of the local 'poll' tax to be funded and sustained. In addition, the much smaller political élites present in these regional capitals have been able to work across departmental boundaries at all levels with considerable ease, away from the gaze of Westminster, shielded from much public accountability, and motivated to work for the common good of their sub-national territory.

This configuration of administrative devolution and regional development agencies working successfully in Scotland, Wales and Northern Ireland has naturally attracted the envious eyes of some English regions. In the 1970s, some of the bigger, quasi-regional authorities set up their own regional enterprise boards with funds from their own employees pension funds, county council grants and some private finance. The West Midlands and Greater London Enterprise Boards even managed to survive – but in much truncated form – the demise of their founding local authorities. Lancashire Enterprises Limited had acquired a national reputation for innovative and enlightened investment and help for local firms. But the aversion to regional approaches in England had been mirrored at this level too with individual local authorities each embarking on their own local economic initiatives, with rare examples of cooperation (such as between Devon and Cornwall, or Yorkshire and Humberside) serving as exceptions that prove the general rule. Some English county councils have argued that they are sufficiently large to count as regions in their own right – Essex (population 1.25 million) is one such council to have pursued this line, even to the point of twinning with the French region of Picardy and setting up a joint representative office in Brussels. At least eight British regions, counties or districts have separate representatives in Brussels, alongside Scotland, Wales and Northern Ireland (Audit Commission, 1991).

By the end of 1993, two further developments in the organisation of local administration point in contrary directions as regards the future of the regional dimension. On the one hand local government reform was put in hand with the Local Government Commission under Sir John Banham being directed to establish unitary local authorities, often at the expense of the long established county councils. However, local opposition all over England meant that this plan was only able to be implemented in part. The Banham Commission did not even ask respondents in its surveys of public opinion whether they had any sense of regional identity, unlike the Redcliffe-Maud and Kilbrandon commissions (Royal Commission on Local Government in England, 1969 and Royal Commission on the Constitution, 1973). In Scotland, a more radical scheme abolished all regional councils with effect from 1994. Yet in an attempt by central government to streamline the organisation of central government departments at regional level, Whitehall has proposed to establish new regional directors at a senior level to coordinate the work in the standard regions of several Government departments (regional economic development, environment, employment and training, industry and transport) and to offer a 'one-stop' shop to firms and local authorities (White, 1993a and Willman and Burt, 1993). This may indeed be a classic example of top-down regionalism, of devolution rather than decentralisation, but it is almost as if the very fragmentation of local government at base level was driving the Conservative Government inexorably towards the recognition of the region for strategic planning and development purposes, in defiance of its own rhetoric, although this may have been at the expense of elected and directly accountable local government (Capon, 1993).

The European Community Dimension

Regional policy has become a major concern of the European Community since the late 1960s when European Community leaders from the six original Member States realised that a strong Community regional policy was an essential accompaniment to the project for economic and monetary union. This interest in regional policy, which was of most concern to Italy, was then exploited strongly in the enlargement negotiations of 1970-1 when the United Kingdom and Ireland, as applicants to join the EC, sought and obtained compensation for their peripheral position relative to the main Community markets. Britain was also keen to establish a budget line, in the form of the new European Regional Development Fund (ERDF), from which it could draw large amounts (initially 28 per cent of the entire Fund) in order to balance some of its potentially large 'net contribution' to the overall European Community budget. By the mid-1980s the ERDF had expanded considerably on the back of further additions to the membership of the Community and was increasingly being organised in combination with the European Social Fund (ESF) and the guidance section of the fund financing the whole of the common agriculture policy. The three funds, the so-called structural funds, will account for well over a third of European Community budgetary spending by the year 2000, compared with less than ten per cent of spending in 1980.

A more long-standing constraint on the development of national regional policies has been the Community's control over state aids, using Articles 92 to 94 of the European Community Treaty. The logic of European Community competition policy demands that national governments should not be permitted to distort competition and subsidise 'lame duck' enterprises which are otherwise uncompetitive and not viable. The European Commission (DGIV) in Brussels must therefore authorise any regional development scheme, using state subsidies, in any Member State, and any subsidy to a particular sector of industry or to a particular firm. The Commission has tried to make a judgement on the 'objective need' of particular regions using its own consistently comparable data of regional deprivation and divergence, and establishing a sliding scale of permitted levels of subsidy (in terms of percentage of the whole capital cost of a project) applicable to each region of the Community. While the controls on state aids from Brussels were honoured as much in the breach as in the observance in the 1970s, the drive to complete the single European market begun in 1985 has encouraged successive competition commissioners to apply the rules very comprehensively and to require much more information about, and control over, regional and sub-regional incentive schemes to encourage existing and newly-formed businesses to grow. Although the Commission is permitted to allow subsidies where conditions justify them, this detailed control has nevertheless begun to bite, and contributes significantly to the way United Kingdom government departments, local authorities and enterprise agencies can structure their economic incentive packages for regions and areas in difficulty.

In the United Kingdom the process of European integration has thus heightened the role of the Community in the management of regional development at the level of Whitehall, and below. Securing eligibility for receipt of structural funds from Brussels, and then competing for those funds once eligibility is established, has proved to be a major strategic consideration for regions as varied in their needs as the West Midlands, Strathclyde, and Devon and Cornwall. As state aids have become more heavily controlled, and fewer job creating investment opportunities presented themselves, so the bulk of European Community funding has been more and more devoted to the co-finance of infrastructure projects. Major Whitehall constraints on local authority capital spending have made it increasingly difficult for EC-funded regional schemes to be launched, and there are long-standing disputes between Brussels, Whitehall and local agencies over the extent to which central government pockets money received from the Community rather than passing it down the line – this is the issue of 'additionality', over which the Commission and other Member States have made periodic threats to withhold funds from Britain.

These sums received from the structural funds are not inconsiderable – around £900 million per annum, in the late 1980s, nearer £700 million each year in the 1990s. They have acquired extra significance for the regions as United Kingdom government spending on regional development grants was cut back sharply in the 1980s. The result has been that regional spending priorities have been increasingly dominated by European Community, not United Kingdom, eligibility criteria and policy objectives, with the Department of Trade and Industry, the lead department in Whitehall, content to travel along the path toward greater Europeanisation of regional policy (Department of Trade and Industry, 1988). Local and regional authorities' bids for assisted area status, a *sine qua non* for receiving ERDF support, have increasingly had to be made to

Brussels as well as to London. Whereas in the mid-1980s only Northern Ireland was deemed by the European Community to be worthy of top priority funding from the structural funds, with mid-Wales and Devon and Cornwall obtaining some rural development help, by 1993 Merseyside and the Highlands and Islands were added to the list of 'objective one' regions, and towns in South-East England, such as Portsmouth and Margate, were bidding to be included.

EC funding for poorer regions, concentrating mainly upon capital projects for infrastructure and industrial development and vocational training schemes, demands much new activity for United Kingdom central and local government, especially as England was stripped of almost all regional administration by the Thatcher government. Regional development plans have had to be submitted to Brussels on the basis of which Community Support Frameworks, directing European Community aid, have been negotiated. Funding from Brussels for integrated operations covering areas in and around Belfast, Birmingham and Glasgow has been conditional upon a regional team of administrators and local representatives being formed to implement and monitor the approved major funding programmes. Regional GDP statistics have, very controversially, had to be prepared in addition to regional unemployment statistics, so that Brussels can take a pan-European view of Britain's regional needs. In the absence of any defined regions in England (other than the over-large eight 'standard' regions), county councils have had to bid for regional status in the eyes of Brussels, or make alliances with their neighbours to do so. Local authorities have had to become increasingly wise to the policies and politics of the European Community, in order to benefit from the myriad programmes, with subsidies attached, emanating from Brussels and in order to comply with and implement, as agents of central government, the letter and the spirit of the Single European Market. They have been instrumental in alerting the Commission to the specific problems of regions facing industrial shutdowns in coal, shipbuilding, steel, defence and other sectors, prompting in turn European Community-led targeted conversion programmes such as RECHAR, RENAVAL, RESIDER and KONVER respectively. The Commission has seemed to many in local and regional government, for a long time, to be more sympathetic and more in tune with regional needs than Whitehall.

It is not just that the whole administrative structure of England is unsuited to the structure for developing and administering European Community structural fund programmes (the position in Scotland, Wales and Northern Ireland being wholly different) which the Community has laid down for all Member States. England lacks a regional tier of administration, and the small regional outposts of central government departments have not been in the habit of cooperating one with another. The Wilsonian structure of regional economic planning boards and councils from the 1960s was finally dismantled in the early 1980s. Strategic urban planning authorities, such as the Greater London Council and the metropolitan country councils were all broken up in the mid-1980s and not replaced. This left no apparatus in England with which to handle the ever-growing need of Brussels for regional interlocutors and regional economic analysis, as was pointed out by a House of Lords Select Committee in 1984 (House of Lords Select Committee on the European Communities, 1983-4 and 1987-8).

By the autumn of 1993, the United Kingdom government was forced publicly to announce a reversal of policy and a strengthening of the regional offices of central government departments such as the environment and employment ministries (White, 1993b) – the purpose of these reforms being ostensibly to exercise greater control over the urban development corporations and to enable civil servants to respond better to European initiatives.

Another event in 1993 tellingly illustrated how the process of centralisation of power in the British state was colliding with developments in the European Community. The Maastricht Treaty on European Union provides for the creation of a European Community-wide Committee of the Regions to advise the European Community institutions on regional needs and issues. The United Kingdom government did not oppose the creation of this new quango, having other more important fish to fry in the treaty negotiations, but it proved surprisingly resistant to demands from the House of Commons that all 24 United Kingdom representatives should be drawn from elected local government councillors. It appears that the Conservative Government would have liked to put British ministers with regional portfolios on their Committee, but MPs – perhaps fighting other battles – decided to insist on a councillors-only clause, and the government was defeated.

The Region in British Society

Although British society is relatively homogeneous in relation to other European states there are still many factors in its composition which make for diversity, many of which are manifested in differences in the geographical distribution of characteristics, lifestyles and traditions. These differences form much of the basis of regional sentiment and identity in the United Kingdom, as well as the fact of geography itself.

The British population itself has diverse origins. While very few can identify their Norman ancestry, let alone their origins as incoming Jutes and Angles from continental Europe, there is still a keen sense of the different ethnic identities in the Celts in Ireland, the Picts and Scots in Scotland, the Norse people in Orkney and Shetland, and the mixed Celtic-Briton population of Wales. More recent migrations of Irish, Afro-Caribbean and people from the Indian sub-continent have not taken on a regional form but particular urban concentrations. This means that some ethnic differences in Britain have a regional cultural resonance while others do not. This is seen in terms of differences of first language, for example, where the large Urdu, Hindi and Gujurati speaking populations are spread across several urban centres in England, while the smaller population of Welsh speakers (*circa* 500,000) is concentrated in Wales, with especially high concentrations in the north and west of the country. Minority languages specific to the British Isles (Gaelic in kind) also occur in Ireland and northern Scotland. Dialects and the incidence of particular idioms and word forms also have a strong regional distribution throughout Britain as elsewhere.

Popular culture is also subject to regional variations, despite the pervasive influence of nationally organised mass media. Television has important regional opt-outs (especially for Scotland, Wales and Northern Ireland) as regards the BBC, and independent television, apart from its national news service, is completely regional in its organisation. A long campaign led to the creation of a separate largely Welsh medium television channel in 1980 (S4C) subsidised by the whole independent television network. The daily press is even more biased towards London production and orientation, but does find it difficult selling papers in Scotland and Northern Ireland, where indigenous dailies such as *The Scotsman* or *Belfast Newsletter* or regional variations of the national dailies are more favoured. The arts in general have developed an increasing regional profile, with national orchestras, theatre and dance companies making regular touring appearances in regional centres, and the regional centres themselves developing a substantial artistic life. Part of the Tate Gallery's collection in London has been moved to new permanent sites in Liverpool and St. Ives (Cornwall), while Glasgow has made great strides to develop a cultural profile that rivals that of Edinburgh. Leeds now hosts an internationally recognised piano festival and Birmingham has a symphony orchestra of international standard. Market researchers have found important differences of taste occurring between regions whether it be fashion or food and drink. The Welsh have a penchant for tinned salmon and the Geordies for brown ale, it seems. In sport, the more thriving activities often have regional organisations to implement national rules, as a practical response to supervising events and behaviour of athletes rather than a real expression of regional identity (the Welsh Rugby Football Union excepted!). Some sports do have a particular regional presence, rugby league in northern England, hurling in Ireland, and Highland games in northern Scotland. The voluntary sector also spawns regional organisations as a necessary interface between national and local activities, the more extensive its area of work and activities become. A few trade unions are region-specific such as the Educational Institute of Scotland and the Farmers Union of Wales, while in Ireland trade unions often ignore the border for organisational purposes.

In the world of education there are pronounced differences between Scotland, Wales, Northern Ireland, and the rest of Britain. Thus the break between Scottish secondary and higher education occurs one year earlier than in England and Wales, and a different examination system and degree course structure results. In Northern Ireland almost all schools are still divided on communal lines, while in Wales, the Welsh language has a special place in the national curriculum and numerous Welsh medium primary and secondary schools and pre-school groups have been set up since the 1950s. The University of Wales has a special federal structure, but this appears to be breaking down (rather as is that of the University of London).

Variations in religious affiliations and observance are also to be found in the regions within the indigenous Christian tradition. Thus church-going in Northern Ireland is a majority pursuit, compared to one in seven of the United Kingdom population as a whole. Important Catholic concentrations occur in the West of Scotland, Merseyside and other areas of high Irish immigration since the 1840s. Non-conformist Protestants are strongly concentrated in the South West and North of England, Wales and Northern Ireland. In Scotland a more advanced form of Protestantism prevails, and the Church of Scotland has more than once provided the focus for national campaigns on behalf of the whole nation. The Anglican tradition provides the established state church only in England.

Regional religious differences form part of the reason for regional political variations in voter allegiance. Catholics provide disproportionate support for Labour in central Scotland, and the non-conformist past accounts for a large part of the survival of the Liberal tradition in the West of England. In recent years general elections have demonstrated more pronounced variations by region in the swings of voter support for political parties. The 1987 election saw an ever-sharper loss of support for the Conservatives the greater the distance from London, but a favourable swing to the government in the South-East of England (see Butler and Kavanagh, 1988, p. 284). In 1992 the reverse occurred, with the largest swings against the government occurring in the South-East and the South-West (Butler and Kavanagh, 1992, pp. 324-332). The political battle is somewhat different in both Wales and Scotland where nationalist parties make the party contest a four-way split. In Northern Ireland, none of the British political parties is a significant player in the politics of the province, and the 17 MPs attending Westminster are drawn from four distinct Northern Irish parties. Political differences of this kind have led the United Kingdom parliament to continue to legislate separately for Scotland, with its own Romano-Dutch legal system. Northern Ireland, and occasionally Wales too, also benefits from separate legislation and different standards, where moral and social issues are at stake.

The distribution of public expenditure is also skewed in favour of the more peripheral regions of the United Kingdom. In part this reflects the great role the public sector has played in the economic life of such regions (although this is fast diminishing), and in part it reflects different degrees of economic activity among regional populations. The tax take is much greater in southern England where incomes and employment rates are higher. By the same token, cuts in welfare payment affect disproportionately those living in areas furthest from London (Hencke, 1993). The imposition of VAT on fuel consumption from 1994 bore hardest, for reasons to do with the colder climate, on those living in the north of Britain. The new uniform business rate introduced in 1990 was intended in part to tax businesses in southern England more heavily than those in the north.

As noted earlier, patterns of employment are regionally distinct, and this affects not only the balance between the private and the public sector, but also the growth and popularity of self-employment. Between 1979 and 1987, for example, self-employment in Great Britain grew by some 52 per cent, but the regional growth varied from Wales (+19 per cent) and Scotland (+21 per cent) to South-West England (+90 per cent) (Department of Employment, 1988). In some respects the regional organisation of major public sector employers such as British Coal and British Steel has atrophied as the industries themselves have declined. Regional organisation was abolished in the gas supply industry in the early 1980s, and has been much reduced as a result of the privatisation of British Rail. Regional organisation has however been institutionalised upon the privatisation of the electricity and water supply industries. Regional health authorities overseeing health care provision survived despite government plans to abolish them by 1996, but some regionalisation of police functions has occurred and more is being canvassed (*The Guardian*, 22 October 1993 and *The Independent*, 22 October 1993). A major problem with such regional institutions is that the territorial divisions within which they operate almost never coincide across different functions. Where regional provision of public services, of public infrastructure and some strategic planning is called for the ensuing arrangements have been invariably *ad hoc* in nature (e.g. road building) or nothing has been attempted (e.g. waste disposal strategy).

Economic differences between the United Kingdom regions have also led to striking divergences in the behaviour of the housing market between the regions in the 1980s and 1990s, the boom of the former especially in the south and east being followed by a major bust. Because home ownership using mortgage finance affects half the population, major differences in purchasing power have emerged between the regions as regards non-housing related expenditure. This in turn has altered the severity and impact of the recession across the regions of the United Kingdom and has contributed to the apparent convergence of regional economic indicators noted above.

The portrait of the United Kingdom regions thus depicted is confusing, as regional variations in social, economic, political and cultural life are revealed to have many different, almost kaleidoscopic, configurations. Yet the particular distinctiveness of Scotland, Wales and Northern Ireland emerges clearly and this has been recognised in the arrangement of public administration, and to a lesser extent the law, of these countries. The position in England is less clear-cut, for while regional differences do exist they are by no means so sharply drawn compared with the rest of the United Kingdom, nor do their contours very often overlap. There is no regional administration or government or public life to give shape or form to such regional differences as do exist, or to build up a regional constituency of shared interests. Indeed in many parts of middle and southern England no clear regional identity exists among the population at large. Where some regionalisation of government was beginning to emerge, and an identification with it by public opinion, the institutions were destroyed in 1986, in the case of Greater London much against the views of the electorate. A contributory factor, in addition, has been the much stronger local identities, typically to be found in counties such as Somerset, Cornwall or Durham, which reflect administrative (and socio-cultural) divisions which go back 1,000 years. The French *départements* (200 years old) and the German *Länder* (less than 50 years old) cannot compete. Changes in communications are making for a reconfiguration of identities and interests. Nationally dominated television, radio and newspapers have spread national perspectives and values to the whole population of Britain, and have contributed to the further marginalisation of regionally-based ethnic cultures, provoking a spirited political response. One of the most dramatic of such responses was the hunger strike started by Mr Gwynfor Evans, former Plaid Cymru leader and MP, in 1980 to persuade the new Conservative government to honour its election commitment to establish a separate Welsh fourth television channel. The government succumbed to pressure on this issue.

The shrinking of the separation of the periphery from the core of the United Kingdom and its capital as a result of high-speed rail and air travel, and the development of an extensive motorway network, has opened up many previously remote parts of Britain to outside influences and has made the notion of regional government and regional identification more feasible. The great paradox is that just as these possibilities are being opened up, the official position of the United Kingdom government until 1997 could scarcely have been more hostile. Indeed one construction to be put upon the administrative and structural reorganisation of government and the public sector pursued since 1979 is that the Conservatives have consciously sought to undermine all large centres of power located away from Westminster with a view to reinforcing the power of central government. A parallel rhetoric can be deployed to claim that the break-up of big local authorities and the waves of privatisation will deliver more local control to consumers, customers and communities. But their margin of control has been reduced by Whitehall strings attached, and by reductions of real capital and real current spending limits. In relative terms central government has won back power from the regions and has ensured that no regional institutions in England with any political weight have survived to challenge Whitehall's priorities. The final irony appears to be that, in response to a mixture of excessive fragmentation of local government and European Community pressures, central government itself had to re-invent the regional tier of administration, albeit initially completely subject to Whitehall control.

Conclusion: Why is Britain the Odd One Out in Europe?

Regionalism was always likely to have difficulty winning political acceptance in the United Kingdom. It appeals to planners and to the politically marginalised, especially those on Britain's periphery, but the idea of regionalism does not sit well within the political culture of the 20th century British state. Since the Norman conquest in the late 11th century the ideology of firm central government control over the regions as an essential condition for the survival of the state and the exercise of British power elsewhere in the world has been dominant. All opposition to those in charge of the united British State has been neutralised, whether external or internal in origin, often by use of 'divide and rule' tactics. The unionist ideology, buttressed by pragmatic arguments, has been supported by appeals to resist incursions on British sovereignty by Brussels and by outbreaks of jingoism and reinforcement of national pride occasioned by victory in world wars and the 'successful' Falklands war. The national tabloid press, which is very extensively read, has been a willing accomplice in reinforcing such attitudes.

There is an implicit assumption too that a weakening of the bonds that tie the components of the British state together will work to the advantage of Britain's more powerful competitors, undermining British effectiveness on the world stage. According to this argument, a strong Britain is better equipped to resist any Franco-German conspiracy in Brussels or the escapades of any United States administration. Britain, despite its much reduced economic and political weight, must still have its own independent nuclear defence capability, larger than average standing armed forces, and its own seat on the United Nations Security Council. Applying this perspective, there is no real recognition that other large states have sought to and succeeded in drawing strength from the institutionalisation of their own internal regional diversity and needs.

Unionist ideology has combined happily with a mistrust of local government

	1971	1979	1982	1986	1988	1993	1996
England							
South East	2.2	2.7	7.0	8.5	5.2	8.6	5.4
South West	3.3	4.2	7.9	9.7	6.3	9.5	6.2
Eastern	3.1	3.3	7.7	8.8	4.8	9.4	6.1
East Midlands	2.9	3.5	8.8	10.3	7.3	9.5	6.8
West Midlands	2.9	4.2	12.2	12.9	8.5	10.8	7.4
Yorkshire &							
Humber	3.7	4.4	10.8	12.8	9.5	10.2	8.0
North West	3.8	5.4	12.4	14.4	10.6	10.6	8.0
North East	5.5	6.9	13.6	15.7	11.8	12.9	10.6
Wales	4.3	5.7	12.4	14.4	10.3	10.3	8.2
Scotland	5.7	6.2	11.7	13.8	11.4	9.7	7.9
Northern Ireland	7.1	8.1	14.7	18.3	16.3	13.7	10.9
All of the UK	**2.8**	**4.3**	**9.9**	**11.5**	**8.0**	**10.3**	**7.5**

Table 1: Regional unemployment rates 1971-1996 (%)
Sources: Department of Employment and Office for National Statistics

leadership and its integrity. Although all too many Westminster politicians have 'fallen short' in their personal lives or embarked upon foolish policy nostrums (the 'poll tax' fiasco of 1988-91 being a fine example), it is the vicissitudes of certain local councils (especially in Labour parts of London) and of fallen heroes such as Newcastle's T. Dan Smith that are best remembered. The British political élite (like the Irish political class) do not trust local government and local politicians, an attitude which conveniently feeds the trend towards even greater centralisation of power in London. Regional government and elected regional assemblies have been depicted as introducing extra and unnecessary layers of bureaucracy. Arguments for greater accountability over the exercise of executive discretion were brushed aside. There has until very recently been a pervasive attitude of fatalism about the efficacy of any reform of United Kingdom political institutions, not so surprising in view of the widely-perceived 'failures' of all health service and local government reforms since the 1960s.

Ideology and scepticism about the politics of regionalism have also found support from strong localist sentiment especially in parts of England, which is focused on the traditional counties, with their long history and tradition. The county councils may well prove to be the casualties of the reforms of government now gathering pace in England and as regional economic institutions become established and legitimised once more.

Another unspoken factor at work is the fear felt by the Westminster-based political class concerning the creation of rival power bases and potential competitors in the political process. This was most graphically illustrated by the refusal of the Houses of

Parliament to give any privileged access to its premises to the 87 elected members of the European Parliament for the first 11 years of their existence. The MEPs were seen as poaching on the constituency work of MPs and were, it seems, not to be encouraged or assisted in their work. Westminster has much the same self-interested approach to the possibility of delegating its work to and offering an alternative significant political career structure through new regional political institutions.

Although much of the above analysis has shown that until 1997 there were few grounds for optimism about the prospects of regionalism taking root in England, the inescapable pressures within the United Kingdom political system, starting with demands for Scottish devolution, have now put the idea firmly into the area of practical politics. The English dimension to devolution has been a critical part of the broader ideological argument on devolution which has sought to counter notions of merely pandering to nationalism by offering some form of assembly. The European Union dimension has also played its part as a lever in the process of satisfying a demand for some devolution all round, with representatives of local government increasingly of the view that stronger regional tiers of government to be found among continental European states have real advantages in comparison with the centralised British state. There has at last come to be a progressive centre-left consensus that the unitary political structure of the United Kingdom bears a large responsibility for its lack of political and economic success since 1945 (Hutton 1995; Marr 1996). Yet the general public in England and Wales still needs some persuading that regional government really will make a difference and will not just be another tier of bureaucracy paid for by the taxpayer. The case for more accountability to democratically constituted authorities gets stronger as administrative regional structures are strengthened. The government offices in the English regions are proving to be significant players in regional economic development. The new regional development agencies which are being set up will add a further dimension to the accountability argument. The slow development of local government pressure for a regional tier of government has now accelerated rapidly with the proposal from central government to establish representative 'regional chambers' throughout England.

Of course the installation of regional institutions of government all over the United Kingdom will not all be plain sailing nor, given the different systems for their composition, are they likely to be mere mini-Westminsters in their behaviour. Friction between the different levels of government will be inevitable and a new forum for negotiations between Whitehall and the regions will be needed. The regions are already building up links with the European Union either by direct representation in Brussels (Greenwood 1997) or through the official Committee of the Regions. A reformed House of Lords could easily offer regional interests more political influence in Westminster. Sooner than many had dared to imagine the political class itself is having to decide, each individual for himself or herself, whether to opt for a career in national politics at Westminster or to choose to develop the new systems of governance at Scottish, Welsh or English regional level. The political heavyweights are making different career choices for the first time in living political memory – a sure sign that regionalism in the United Kingdom is here to stay and another indication that, reluctantly and after everyone else, the United Kingdom is coming slowly into line with its European Union partners on this aspect of government as in so many others.

References:

Audit Commission (1991), *A Rough Guide to Europe: Local Authorities and the EC*, London, HMSO.

Bogdanor, V. (1979), *Devolution*, Harmondsworth, Penguin.

Butler, D. & D. Kavanagh (1988), *The British General Election of 1987*, London, Macmillan.

— (1992), *The British General Election of 1992*, London, Macmillan.

Butt Philip, A. (1975), *The Welsh Question*, Cardiff, University of Wales Press.

— (1978), *Creating New Jobs*, London, Policy Studies Institute.

Capon, B, *The Times*, 2 November 1993.

Department of Employment, *Preliminary Results from the 1987 Labour Force Survey* (1988), Table 4.

Department of Trade and Industry, *DTI – The Department of Enterprise*, London HMSO, Cmnd. 278, January 1988.

Greenwood, J.(1997), *Representing interests in the European Union*, New York, St Martin's Press.

Hanham, H. (1969), *Scottish Nationalism*, London, Faber.

Hencke, D. 'Welfare cuts hit poorest areas worst', *The Guardian*, 20 September 1993.

House of Lords Select Committee on the European Communities, *European Regional Development Fund*, 23rd Report of Session 1983-84, HL Paper 274, pp. xxix-xxxii.

House of Lords Select Committee on the European Communities, *Reform of the Structural Funds*, 14th Session 1987-88, HL Paper 82, pp. 18-19.

Hutton, W. (1995), *The State we're in*, London, Cape.

Marr, A. (1996), *Ruling Britannia: the failure and future of British democracy*, London, Penguin.

Moores, B., J. Rhodes & P. Tyler (1986), *The Effects of Government Regional Policy*, London, Department of Trade and Industry, HMSO.

Report of the Royal Commission on Local Government in England (1969), Cmnd. 4040, London, HMSO.

Report of the Royal Commission on the Constitution (1973), Vols I and II, Cmnd. 5460, London, HMSO.

Rose, R. (1982), *Understanding the United Kingdom*, Harlow, Longman.

Smith, D. (1989), *North and South: Britain's Growing Divide*, Harmondsworth, Penguin.

Tighe, C., 'Portsmouth to compete for EC aid', *Financial Times*, 12 October 1983.

White, M. (1993a), 'Confusion on urban renewal supremos', *The Guardian*, 15 October 1993.

— (1993b), 'Regions regain planning clout', *The Guardian*, 5 November 1993.

Willman J. & T. Burt, 'Business help sought in urban revival plan', *Financial Times*, 1 November 1993.

Wintour, P., 'Whitehall seeks EC aid for depressed South-East', *The Guardian*, 12 October 1993.

Regionalism in the Republic of Ireland

Ullrich Kockel

Introduction

Although the Republic of Ireland is supposedly a capitalistic market economy, the highly centralised state has played an extraordinarily strong part in economic development during this century. This has often been linked to the country's colonial past (Fennell, 1983). In the circumstances, it is hardly surprising that the regional dimension of Ireland has remained somewhat underdeveloped, although a degree of rather tentative regionalisation has been achieved in recent years under pressure from the European Union. More or less realistic proposals for cultural and politico-economic regionalism have surfaced periodically since the 1960s, but in practice little progress has been made. Curiously, the spatial structure and hierarchy of Ireland seems to have emerged along with, and to this day is following, the arterial railway network of the United Kingdom with its centre in London. The Free State after 1921, and the Republic after 1937, failed to challenge this spatial imbalance for what it was and still is – the most visible reminder of Ireland's economic dependency on Britain.

The Colonisation of the Regions

Before the Act of Union in 1800, Ireland was never a unitary nor indeed, despite nationalist claims to the contrary, a unified country. In early medieval times, the office of a High King existed, but appears to have been largely a ceremonial (and frequently contested) one. As elsewhere in Europe, real power lay with local 'kings' of whom there were some 150, and with their immediate superiors, the provincial kings of Connacht, Leinster, Munster and Ulster. Since the Norman conquest in the 12th century, this provincial division of the country has gradually lost all real political relevance. At the cultural level, the provinces have retained a function as an intermediate tier in national competitions. The treatment of Ulster, in this context, is interesting. Where an activity is organised on an all-island basis, that is, covering the 'national territory' according to the Irish Constitution of 1937, the nine counties of Ulster compete at the provincial level regardless of the state border dividing the province. If the activity is organised in the Republic of Ireland only, the three southern counties of Ulster are usually grouped with Connacht.

Outside the field of sports and competitive cultural activities like Irish dancing, there is no significant role for the provinces. Loyalties are primarily local, and beyond that mainly expressed at the county level. The county is also the effective level of such local government as does exist. Established in the late Victorian period as part of a wider local government reform in the United Kingdom, some counties were broadly based on earlier political entities, but most were new constructs serving the administrative needs of the state that designed them. Older territories that remained largely intact, like the 'Kingdom of Kerry', retained a strong regional identity, giving rise to cultural coping strategies in the rest of the country – in this case the 'Kerryman' jokes – which in turn reinforced the region's distinctive identity status.

Outside the Dublin area, such identities are particularly strong on the western seaboard. The West of Ireland is widely perceived as the most peripheral part of the island, making it one of the most peripheral areas of the European Union. However, this was not always the case. At the beginning of Europe's expansion overseas in the 16th century, the city of Galway in Connacht was for a time the second port of the British Isles, maintaining strong trading links with Spain, the Mediterranean and North Africa. In the south-west, the city of Cork in Munster was an important commercial centre for the continental and transatlantic trade. At the time Dublin, in Leinster, was still relatively small, and Belfast, in Ulster, founded only in the 17th century, did not even exist. From this initial position, a much more balanced regional system might have emerged. However, during the 17th and 18th centuries the colonial administration succeeded in shifting the balance of power and wealth in the island further towards the east. A detailed analysis of this process cannot be attempted here, but it should be noted that a major impetus for this policy was the fear that Ireland might be used as a 'back door' for invasion by the rival colonial powers, Spain in the 17th and France in the 18th century. Moreover, with the Industrial Revolution, which concentrated industrial potential in the north-east and destroyed much of the cottage industry elsewhere in the island (Walsh 1995, p.54), Ireland became a chief source of cheap food for Britain's industrial heartland. Agricultural exports were largely redirected to Britain and to the new economic core area in east Ulster, and the old trading centres lost much of their business. Dublin had long been the administrative centre of the island, and with its enhanced role as the main port for exports to Britain began to prosper and expand. The development of the railway system accelerated the growth of the city (Walsh, 1995) that increasingly came to regard the rest of the island as its hinterland. By the time 26 counties separated from the United Kingdom to form the Irish Free State, the reality of Dublin dominating the new country in every respect was well established, and the new government was too busy building the state to do much about it.

From Free State to Free Market

Unlike other European countries, the political divide between the major parliamentary parties in Ireland is not between Left and Right, but between the pro- and anti-Treaty parties of the Civil War – between those who then accepted partition of the island (Fine Gael), and those who did not (Fianna Fáil). From the 1920s to the 1950s, the dominant political ideology in the Free State and the Republic was one of self-sufficiency. Ironically, the pro-Treaty government during the first decade of independence failed to recognise that with partition, the state had been deprived of an indigenous industrial

IRELAND

DONEGAL

NORTHERN
IRELAND

LEITRIM

SLIGO

MONAGHAN

MAYO

BORDER

CAVAN

LOUTH

WEST

ROSCOMMON

LONGFORD

MEATH

MIDLANDS

MID-
EAST

WESTMEATH

GALWAY

OFFALY

KILDARE

Dublin

DUBLIN

LAOIS

WICKLOW

MID-WEST

CLARE

TIPPERARY N

CARLOW

KILKENNY

LIMERICK

SOUTHEAST

TIPPERARY S

WEXFORD

WATERFORD

SOUTHWEST

KERRY

CORK

Gaeltacht areas
Designated areas

0   50   100

kilometres

base. While elsewhere in Europe the new small states tried to create jobs by industrialisation, the Irish government saw the economy as depending on agriculture, and sought to raise general prosperity by promoting agricultural exports (Kennedy, Giblin and McHugh 1988, p.34). The economist Keynes endorsed this view in a famous lecture at University College Dublin in 1933. By that time, however, the 'anti-Treaty' party had gained power. An economic war had broken out between Ireland and its main market for agricultural exports, Britain, after the new government withheld annuities arising from the purchase of land by tenants prior to 1921, which had been funded by the British government. In retaliation, Britain imposed a levy on Irish agricultural imports. The war escalated, lasting until 1938, and provided the rationale for the protectionist policies pursued by the Irish government.

The year 1958, when the first Programme for Economic Expansion (1959-63), was published, is generally regarded as a turning point in Irish development policy (Brunt, 1988). The protectionist measures that had accompanied slow development for two decades were lifted because a home market, shrinking due to high emigration, held little promise of job creation and general growth. Foreign capital, which had been viewed with suspicion after independence, was now invited into the country – the road was paved for export-led growth strategies that have dominated Irish policy thinking ever since. When, in 1973, Ireland joined the European Community with its emphasis on free trade, this trend became even more pronounced. In 1982, the government-sponsored Telesis Report pointed to 'deeprooted structural problems caused by dependence on foreign investment' (Munck 1993, p.156), highlighting foreign firms' lack of linkages with the rest of the economy, and the need for a dynamic indigenous industrial infrastructure. This was echoed in the Culliton Report ten years later, indicating that little had changed in the meantime.

During the 1970s, after more than a century of continuous population decline, Ireland experienced substantial population growth as a result of high birth rates and net immigration. The 1981 Census reported 55.6 per cent of the population living in 'aggregate town areas'. The Dublin region in particular has grown dramatically throughout this century. Although there is an undercurrent trend towards rural resettlement, especially among immigrants, the overall tendency is still towards growth in urban areas, and the decline of rural service provision induced by the rural exodus is hardly being checked. Galway Bay, north-west Donegal, and, to a much lesser extent, north Sligo/Leitrim have experienced rapid growth of suburbanisation, as an increasing number of the urban middle-classes choose to live in a presumably unspoilt rural environment. The corresponding expansion of commuting has led to traffic congestion in Galway and Sligo, while in the Gweedore area in Donegal, suburbia has grown largely around the industrial estate at Bunbeg without any urban centre.

During Ireland's early years in the European Community, the government made two controversial choices. The first was to have Irish – in constitutional terms at least the first language of the state – recognised as an official, but not a working language of the European Community. While this decision was taken with a view to the Northern Ireland situation, it created difficulties that are only being recognised now, in the wake of the recent revival of the language. The second decision, which caused a considerable stir at the time and even greater problems since, was to have the Republic classed as a single region for Community Regional Policy (CRP) purposes since 1975, thus ensuring that the centre, Dublin, retained full control over European funds allocated to Ireland. The interests of regions and localities were seen as being best served by this central dirigism (Kockel, 1993). However, the regional dimension of planning was in practice all but ignored, as the 'aspatial nature of the Programme for National Recovery' (Brunt, 1993, p.35) of 1987 demonstrated. The promotion of national development has, in general, been accorded higher priority than regional interests, especially in the formative years of the state and during the recession of the 1980s (Brunt, 1988). At times, governments have 'genuflected' (Ó Tuathaigh, 1986, p.127) towards decentralisation, but this has been characterised by Barrington (1982, p.106) as 'dropping fragments of public bodies in provincial towns', with development being 'done *to* a region by benign central authority, not [...] *by* the region itself'.

The Irish State and its Regions

The Irish system of local government is essentially unchanged since its introduction by the United Kingdom parliament's Local Government (Ireland) Act of 1898. The Act established 27 County Councils (Tipperary was divided into two council areas) and four County Boroughs – Dublin, Cork, Limerick and Waterford. In 1985, Galway also became a County Borough. The Dublin County Council area was divided into three new counties in 1994. There are a number of subsidiary structures, including six borough councils (Clonmel, Drogheda, Dún Laoghaire, Kilkenny, Wexford and Sligo) and 49 urban district councils. The United Kingdom government did not extend the Parish Council system to Ireland, and this meant local development initiative was left to private individuals. A rural social movement, *Muintir na Tíre*, set up parish councils in the 1930s, but these were not incorporated into the statutory structure of local government.

At the cultural level, identification with the county is strong, especially because sporting and other competitions have been organised on a county basis for more than 100 years. However, in contrast to the British system which it was originally part of, the powers of local elected assemblies were significantly curtailed by the Free State's Local Government (Temporary Provisions) Act of 1923, which sought to centralise power and control. The County Manager became *de facto* an agent of central government under the authority of the Department of Local Government (now Department of the Environment). There is no provincial level of government. The Free State also abolished the Congested Districts Board, established in 1891 to address the specific problems of the western seaboard and in this sense an early form of a regional authority. It was not until 30 years later, when the Underdeveloped Areas Act of 1952 created the Designated Areas (see map), that the special needs of the western counties were once again recognised. While the Local Government (Planning and Development) Acts from 1963 onwards appear to give local authorities substantial powers, in practice these powers can rarely be utilised, due to limited resources provided by central government. Prior to 1977, considerable revenue for local authorities derived from local rates, a system of taxation inherited from the United Kingdom. The abolition of domestic rates in 1977 and agricultural rates in 1983 diminished the effective authority of local government even further.

The Health Boards and other government agencies adopted regional structures, some of them following the planning regions established in the wake of the 1963 Act. However, in all these cases, regional structures have remained merely executive agents of a particular central authority, with few devolved powers. Responsibility for economic development, generally perceived as a *sine qua non* of social development, lies with a number of state and semi-state bodies, each with its own regional hierarchy. In 1976, the National Economic and Social Council identified some 200 organisations dealing with different aspects of regional development, the majority of them operating their own regional framework (Brunt, 1988, p.56). Decision-making and control is highly concentrated in Dublin, with any efficient coordination of policies at the local level proving difficult, not least because of the inconsistency of territorial data. The national tourist board, *Bord Fáilte*, has established seven Regional Tourism Organisations, with the Mid-West region being administered by Shannon Development. There are seven regional Fisheries Boards. *Forbairt*, the national authority in charge of technology and entrepreneurial development, operates nine regional offices it inherited when it took over some of the functions of the Industrial Development Authority (IDA). The IDA no longer has a regional structure. The Training and Employment Agency, *FÁS*, is organised in ten regions, three of which are in the Dublin area, while *Teagasc*, the Agricultural and Food Development Authority, works on a county-basis.

The only genuine regional agency, the Shannon Free Airport Development Company (SFADCo), commonly known as 'Shannon Development', is in charge of industrial and tourism development in the mid-West region centred on Limerick. It started in the 1950s as a local agency to promote Shannon Airport and its duty free industrial zone, but soon embarked on a more comprehensive programme of regional development, and over the years has acquired progressively greater responsibility for a growing range of activities. Primarily funded by a grant from central government, it is usually considered the only regional development authority in Ireland. The only other contender for this designation, *Údarás na Gaeltachta*, the authority for the Irish-speaking areas, is in a somewhat different position, as discussed below.

Together with local authorities in the region, Shannon Development created the first Regional Development Organisation (RDO), which provided a model for similar bodies set up in each of the planning regions created after the 1963 Act. The RDO was an informal organisation bringing together representatives of various central government departments and state-sponsored bodies. Following the publication of the Buchanan Report in 1968, central government decreed to establish similar organisations in the other regions. The Buchanan Report advocated a strategy based on a hierarchy of growth centres, a model fashionable at the time, but aroused considerable opposition. Its recommendations were only partly implemented (Walsh, 1995). Instead, the RDOs were introduced primarily to coordinate the work of the plethora of agencies looking after different aspects of development, and to devise regional strategies for the nine planning regions in conjunction with the IDA (Ó Tuathaigh, 1986, p.123). By 1972, the first set of plans was completed. However, with the IDA concentrating mainly on attracting foreign industry and being perceived as having little regard for local or regional issues, some RDOs tried to create a more favourable environment for endogenous development. Although their coordinating role was highly praised (Ó Tuathaigh, 1986, p.124), they were never endowed with executive functions, and were abolished in 1987, possibly because some of them had come to be seen as a threat to the authority of the centre.

In a parallel development, County Development Teams (CDTs) under the auspices of the Department of Finance were set up in 1965 in the western counties classified as Designated Areas, and in West Limerick in 1968. CDTs were chaired by the respective County Manager, and coordinated by a committee in the Department of Finance which comprised representatives of other government departments. Like the RDOs, the CDTs had limited authority. In 1995, their functions were largely taken over by the County Enterprise Boards, established in 1993 to promote local initiative. Following the abolition of the RDOs, Shannon Development had already assumed the functions of the CDTs for Clare and West Limerick in 1988.

Regions with Special Problems

While it can be argued reasonably that the problems faced by Ireland's regions, however severe, are essentially the same as those of structurally similar regions elsewhere in the European Union, there are a number of regions with particular disadvantages. Three of these shall be briefly discussed here.

Dublin

The high degree of centralisation in Irish government and administration has already been noted. Throughout the history of the state, and despite the efforts of the IDA, employment growth in services has fairly consistently outperformed industrial job creation. The bulk of this growth has occurred in Dublin and for much of this century has largely been a result of the expansion of the state. As Ó Tuathaigh (1986, p.129) observed, critics of centralisation argue that 'it draws off vital intellectual and other energies from the [...] regions; encourages a passive or dependency attitude among the 'receivers' [...] of decisions made in Dublin; and [...] encourages a one-way information and culture flow within the state.' The growing polarisation between Dublin and the rest of the country, evident not least in the widening gap in economic indicators, is certainly noticeable (Kockel, 1993). However, '[t]he fruits of rapid urban growth are not all benign, as the experience of Dublin in recent decades has clearly shown' (Ó Tuathaigh, loc.cit.). The accelerated growth in service sector, especially public service, employment, concentrated in the Dublin region, has led to a rapid increase in the population of the area over the past four decades. At the same time, the economic vitality of Dublin's inner city declined as many of the traditional industries closed down. With the containerisation of the port, and the parallel decline in labour intensive port-related industries, employment opportunities were greatly reduced. Persistently high levels of long-term unemployment have created a growing poverty problem in

Dublin (Kockel, 1990). The constant influx of new residents to the region generated pressures for residential developments on the outskirts of Dublin. Several dormitory towns like Tallaght or Ballymun have emerged in recent decades, rapidly overtaking most of the older established cities (except Cork) in size. Dublin has become increasingly ungovernable as a result and, following the Local Government Act of 1991, the county was sub-divided into three new counties – Fingal, Dun Laoghaire/ Rathdown and South Dublin. Each new county has an area manager and will, over a transition period, take over the responsibilities of Dublin County Council.

The Border

While borders and their associated problems are commonplace in the European Union, the Irish case is complicated by the territorial claim to the entire island made in the Irish Constitution of 1937, and vigorously resisted by the Unionist majority in Northern Ireland. A report by the Economic and Social Committee of the European Communities (1983) highlighted the special needs of the Irish border region. Efforts to promote cross-border collaboration have been frequently frustrated not by any practical obstacles, but by local authorities on either side being reluctant to engage with 'the enemy'. The East Border Region (Newry/Mourne and Down districts in Northern Ireland with Louth and Monaghan in the Republic) and a number of other areas have formed cross-border associations based on geographical and cultural homogeneity (McAlinden, 1995, p.81), but others have been less keen. Several Unionist-dominated councils have always regarded cross-border projects as a back door to a united Ireland. A technical problem also arose from the classification of the Republic as a single European Union region. This was interpreted to mean that a project involving, for example, West Cork, as far removed from the border as one can get, could technically be funded under the first Interreg programme. In recognition of this problem, the second Interreg programme for the border, approved in 1995, places greater emphasis on decentralised decision-making. (McAlinden, 1995, p.82).

The Gaeltacht

Ever since the 26 counties assumed their independence, the Irish language has occupied a special position in the state's development policies. Since the mid-1950s, those areas where a large share of the population supposedly uses Irish as their vernacular have been granted special status under successive Gaeltacht Acts. These areas fall almost exclusively within the Designated Areas (see map). Not a coherent region in the strict sense of the term, the *Gaeltacht* are seen as structurally weak rural areas where natural conditions combined with historically evolved land use and settlement patterns, and compounded by unfavourable demographic structures, make conventional development strategies inappropriate. A special agency, *Údarás na Gaeltachta*, is administering state aid for these areas. The title *'Údarás'*, meaning 'authority', is somewhat deceptive – the organisation has a coordinating function, but in many areas such as infrastructure it is frequently overruled by County Councils or other state or semi-state agencies. There have been cases where 20 or more such agencies were involved in a project.

Obvious conflicts exist between language maintenance and a policy of industrialisation and agricultural modernisation. Irish is most widely spoken in rural communities, where modernisation has meant contraction in the number of farms and agricultural employment. Industrial development, together with the decline in agriculture, has brought about social changes that raise doubts over the future of the Irish language. At the same time, the preferential treatment of the *Gaeltacht* has generated conflict in areas that suffer similar or indeed more severe socio-economic difficulties, but do not enjoy the same special status. Additional tension arises from the fact that large parts of the *Gaeltacht* are predominantly English-speaking, especially on the fringes, so that the raison d'être for special treatment is hardly given. Such problems notwithstanding, it was decided that the Gaeltacht should be 'saved', and that the poor people in the West needed to be 'helped' as far as possible. This official attitude may have created the impression among the local population that their areas were lacking valuable resources and initiative, a view which prevailed until the formation of community cooperatives during the late 1960s began to suggest the opposite. Although politicians acknowledged their contribution to regional development, these cooperatives were not made part of any comprehensive, integrated development policy. Another successful 'grassroots' initiative in that period was the Civil Rights Movement, which secured the foundation of an Irish-language radio service in the 1970s. More recently, the establishment of an Irish-language television channel (TnaG), based near Galway, in 1996 has given a further boost not only to the language, but also to a regional resurgence of cultural identity.

The European Union and the Irish Regions

Despite the partial success of some social movements, supported by many intellectuals and substantial research, regionalism in Ireland has been constrained by a national development policy that effectively treats the Republic as a single city region, Dublin, with its vast hinterland. The CRP reforms since the 1980s have forced the Irish government to identify more clearly regional disparities within the country in order to benefit fully from the Structural Funds and the country's Objective 1 status (Brunt 1993, p. 35). Following the Local Government Act of 1991, largely due to pressures from Brussels, local government has been granted somewhat greater powers to pursue of its own accord initiatives deemed to be in the interest of local communities. The European Union, through initiatives like the LEADER programme and more general policy changes, has stimulated a more decentralised approach to development. The old Common Agricultural Policy had resulted in a further widening of the gap between the agriculturally more developed south-east and the less developed west of the country (Ó Tuathaigh, 1986, p.126). In response to this, the LEADER programme seeks to encourage an endogenous development strategy. Intended to operate over relatively large geographical areas with a population of 50-100,000, it has necessitated the creation of regional groupings of interests, varying from relatively small units to entire counties (Walsh and Gillmor, 1993, p.96).

In response to reforms of the Structural Funds, two national plans were produced to secure funding under the Community Support Framework (CSF), covering the periods

1989-93, and 1994-9. The first plan generated a modest measure of devolution. Seven regions were designed to facilitate redistribution of funds under the programme, but instead of signalling commitment to effective regional intervention, these regions showed 'little sign of careful designation' and seemed to 'add to the confusion that surrounds the planning system' (Brunt, 1993, p.36). The plan reflected 'a general disaggregation of [...] *national* sectoral and infrastructural projects onto each of the seven regions rather than [...] the planning agenda of the regions' (*loc.cit.*). However, a process of radical change had been initiated.

In 1994, the planning regions were re-drawn (see map) and regional authorities established, their initial tasks being the management of the Operational Programmes devised as part of the second national development plan, and to prepare regional reports that provide the basis for future plans. The authorities are the first decentralised tier of authority established since 1921. At present, their membership consists mainly of local authority representatives. Whether plans for direct elections to these bodies will materialise remains to be seen. As the Republic is likely to lose Objective 1 status in the foreseeable future, and with it a large proportion of European Union funding, the regional authorities may only have a temporary role. However, the Council of the European Union has recently renewed the mandate of the Committee of the Regions (COR) for a second term. The new Irish regions are represented on the COR, and acutely aware that their future is crucially dependent on the direction in which the Committee's powers develop. For better or worse, Dublin may at last have lost some of its grip on the regions.

References:

Barrington, T. (1982), 'Whatever Happened to Irish Government?', in F. Litton (ed.), *Unequal Achievement: The Irish Experience 1957-1982*, Dublin, Institute of Public Administration, pp. 89-112.

Brunt, B. (1988), *The Republic of Ireland*, London, Paul Chapman.

— (1993), 'Ireland as a Peripheral Region of Europe: Structural Funds and Regional Economic Development', in King, R. (ed.), *Ireland, Europe and the Single Market*, Dublin, Geographical Society of Ireland, pp. 30-43.

Fennell, D. (1983), *The State of the Nation: Ireland since the Sixties*, Dublin, Ward River.

Kennedy, K., T. Giblin & D. McHugh (1988), *The Economic Development of Ireland in the Twentieth Century*, London & New York, Routledge.

Kockel, U. (1990), *Dublin's Inner City: Community-based Initiatives and Employment*, Occasional Papers in Irish Studies 1, University of Liverpool (with M. Ross).

— (1993), *The Gentle Subversion: Informal Economy and Regional Development in the West of Ireland*, Bremen, European Society for Irish Studies.

McAlinden, G. (1995), 'The European Union: A Better Life on the Border', in D'Arcy, M. & T. Dickson (eds), *Border Crossings: Developing Ireland's Island Economy*, Dublin, Gill & Macmillan, pp. 77-84.

Munck, R. (1993), *The Irish Economy: Results and Prospects*, London, Pluto.

Ó Tuathaigh, G. (1986), 'The Regional Dimension', in Kennedy, K. (ed.) *Ireland in Transition: economic and social change since 1960*, Cork & Dublin, Mercier, pp. 120-32.

Walsh, J. (1995a), *Regions in Ireland: A Statistical Profile*, Dublin, Regional Studies Association (Irish Branch).

—(1995b), 'Economic Geography: How Ireland's Wealth is Dispersed', in D'Arcy & Dickson (eds), *Border Crossings*, pp. 53-65.

Walsh, J. & D. Gillmor (1993), 'Rural Ireland and the Common Agricultural Policy', in King, R. (ed.), *Ireland, Europe and the Single Market*, pp. 84-100.

Regionalism in France

Peter Wagstaff

Introduction

The relationship between regionalism and nationalism has long appeared one-sided in France. Political scientists have tended to look upon France as 'the world's consummate nation-state'(Newhouse, 1997, p. 78), an example of 'the state *par excellence*', with little to mediate between the individual citizen and the state (Budd, 1997, p.187). The 'indivisible Republic' has been a commonplace of constitutional texts over two centuries, as successive regimes have sought to affirm and strengthen the integral, seamless nature of the French state. This long-standing desire to achieve as close as possible an identification of state with nation has meant that regional diversity has been neglected, even forced to the margins of French political discourse. And yet this diversity, which has thus been seen as, at best, irrelevant and, at worst, inimical to the health and prosperity of the state, is obvious to even the most casual observer. Considered in terms of geography, climate, demography, culture, language, and economy, France appears as varied as any country in Western Europe. The contrast with the notion of the monolithic 'République indivisible' is inescapable.

Yet alongside this assertion of national integrity, there is ample evidence of thriving regional and local identities, and of the popular tendency to feel most at ease with a local, as opposed to a national, allegiance. Attempts in the 19th century to assess the citizen's feelings for the nation drew so little response, especially in the provinces, that the very notion of France as an entity could seem alien (Zeldin, 1977, p. 3), and more modern and better organised soundings of opinion have generally identified the local (town, *commune*), as opposed to the national, as the focus of a sense of belonging (Lanversin, 1989, p. 78).

There is, then, an underlying and persistent tension between regionalism and nationalism in France. This tension has persisted since the 1789 Revolution, which laid bare the opposition between, on the one hand, the dictates of national unity and the principle of equality and, on the other, the demands of individual, hence local, freedom (Schmidt, 1991, p. 4). Every shift between centralising nationalism and decentralising regionalism that has taken place in France in the last 200 years has reflected that tension. Such shifts as have occurred have been concentrated almost entirely in the last 40 years. Indeed, it could be argued that the only reforms to the administrative structure of France which have seemed likely seriously to threaten centralised national power have been those enacted during the first presidential mandate of François Mitterrand, and specifically in the law of July 1982. While the extent and radical nature of those reforms should not be overstated, they at first took many observers by surprise, and the slow process of consolidation in the years that followed says much about the ambivalence of attitudes among the governing élites.

The decentralising zeal of élites in possession of centralised power was already viewed with scepticism early in the 19th century, when Alexis de Tocqueville declared that only the very disinterested or the very mediocre amongst those in public life wished to decentralise power, and that the former were rare, the latter powerless (Brongniart, 1971, p. 41). A much-reprinted newspaper cartoon by Plantu from the early 1980s is similarly pointed in its satire of such attitudes. The cartoon depicts a peaceable little hamlet, with its church spire and pitched roofs, nestling among rolling hills. On the steps of the tiny *mairie*, the mayor, with his tricolore sash, is asking his deputy 'When's this decentralisation coming, then?' The reply is the sardonic – 'They haven't made up their minds yet in Paris'. The paradox of decentralisation measures which lead to greater regional autonomy being dependent upon decisions taken at the centre provides a telling commentary on the relationship of state and regions in France.

In order to explore the causes and consequences of this paradox and to situate it within the wider context of what François Mitterrand called 'the opening-out of Europe', I shall look in greater detail at the factors which underpin the historic opposition between nationalism and regionalism in France.

These factors fall broadly into two categories – the political and cultural on the one hand, and the economic on the other. The first is the focus for the tension between regionalism and nationalism in ideological terms, while the second encompasses the year-by-year requirements of regional and national economic development. In the historical survey which follows, I shall examine first the political factors which account for the establishment and maintenance of a rigid and hierarchical administrative system. I shall then discuss the economic factors which have called into question the efficacy and even the *raison d'être* of that system. Post-Second World War measures in the political, administrative and economic fields are to a large extent interdependent and will therefore be considered together.

The Centralised State: Historical and Cultural Factors

It can be argued that French mainland territory has been established within natural boundaries for so long that the identity and cohesiveness of the nation are not in doubt. The Atlantic, the English Channel, the Mediterranean, the Alps, the Vosges, and the Pyrenees have produced a hexagon bounded on five of its sides by either sea or mountain range, a land secure within its limits, both homogeneous and diverse in its geography. It was not, however, always so, and the historical succession of conquest, division, war, civil war, leading by a process of accretion, to the achievement of the modern, unified nation state, has left echoes of instability and insecurity which resonate across the centuries. A certain hazy outline of France can be discerned as early as the 9th century, with the dismemberment of the empire of Charlemagne and following the Treaty of Verdun in 843 AD, albeit a France lacking Brittany, Lorraine, Burgundy and Provence. Yet it is only with the accession of Louis XIV, 800 years later, that centuries of acquisition under Louis XI, François I and Henri IV appear

consolidated, and, with the addition of Artois, Alsace, Roussillon, Franche-Comté, followed by Lorraine and Corsica under Louis XV, the map of France at the time of the 1789 Revolution appears much as it is today.

Boundaries are not just political ones, however. Diversity is rooted in culture, in the habits and traditions of a thousand provincial communities, large and small, but above all in the linguistic distinctiveness that marks North from South, periphery from centre. To the extent that there has been a single, hegemonic culture that is French, it is one that is northern, indeed Parisian, with the southern half of the country, not to mention peripheral regions such as Brittany and Alsace relegated, in cultural terms, to a marginality akin to folklore. Braudel has indicated most emphatically that the southern half of France, the civilisation of the *langue d'oc*, with its constituent dialects of languedocien, gascon, provençal, auvergnat, together with the enclaves of basque and catalan, has been subjected to an almost colonial inferiority by the triumphant culture of the *langue d'oïl* in the north (Braudel, 1986, p. 72). The *Édit de Villers-Cotterêts* in 1539 marks the attempt by François I to impose the use of French as the language of the courtroom and indicates an awareness of the importance of French as a unifying influence throughout the kingdom. The establishment of the Collège de France in 1530, where teaching was conducted in French, as opposed to the Latin of the Sorbonne, shows a similar tendency. By 1793 the decision was enforced to conduct all education, throughout the Republic, in French, and from 1880 onwards French children were all guaranteed a secular, public and obligatory education, in French, by state-trained teachers. Up until 1969, most diplomas and qualifications in the secondary and higher education systems were awarded nationally, on the basis of nationally planned programmes of study. The ubiquity of French as the language of officialdom in general, to the detriment of regional languages and *patois*, is the modern concomitant of these measures, and reflects the determination of successive regimes to forge a nation from a range of disparate identities (Battye & Hintze, 1992, p. 18). That determination is expressed in a concept of statehood built around a central source of power which gives form to the expression of national identity, and reaches out to the farthest corners of the territory, imposing its discipline and its authority. It is the Revolution which enshrines these ideas, and gives them concrete form through the new institutions of *départements* and *communes*.

In many ways the reorganisation of France that followed the Revolution set the seal on a process started under the Ancien Régime. Indeed, the proposal for a 'departmental' division was made and rejected some years before the division of the country into 83 *départements* became a reality on 15 January 1790. This division, and the geometrical precision with which it was approached – dimensions of 18 leagues square for the *départements* – had its critics at the time, notably Edmund Burke, who, in 1790, castigated 'the present French power [as] the very first body of citizens, who, having obtained full authority to do with their country what they pleased, have chosen to dissever it in this barbarous manner' (Burke, 1905, p. 149). However, the logic behind the division also predated the legislation – that is the need for a territorial division of such dimensions that a representative of the state could travel, on horseback, from the principal town to any point in the *département* between sunrise and sunset. This marks the first indication that spatial relationships have a role to play in the administrative process, and prefigures the need for different divisions at other times. As Braudel has pointed out, 'the true measure of distance is the speed at which people can travel' (Braudel, 1986, p.48). This simple and obvious fact will determine the rapidly evolving economic geography of France, and therefore the relationship between region and state, throughout the following two centuries.

The aim of the legislators of the Revolution, therefore, was to set in place a system of administration that would underscore the unity of the nation while retaining some sense of historic geographical and cultural realities. The establishment in 1789 of the *commune* (44,000 in total) as the basic unit of administration, and its definition as a 'natural community', whether town, village, or hamlet, echoed these realities, while the *départements* tended to reflect 'natural regions' or old provincial areas, with many boundaries matching diocesan boundaries. The notion of 'natural regions' is, of course, spurious, reflecting a desire to legitimise political decisions about the division of territory by reference to what are in fact nothing more than earlier political decisions on the same subject. Nevertheless, it is with the appointment of Napoleon Bonaparte as First Consul, and the law of 28 Pluviôse, Year VIII (17 February 1800), that we find the appointment in each *département* of a *Préfet* (Prefect), as the representative of central government in whom the power of that government is vested, so that the principal of equality is subordinated to the demands of a hierarchical command structure. This is the origin of the rigid Napoleonic centralisation which characterised French government for much of the next 200 years. Consequently it has been difficult to speak in terms of local *government* in France during this period, since the phrase implies a measure of democratic decision-making and accountability in local affairs. It is more appropriate to refer to local *administration*, that is the means by which the decisions of central government are enforced at local level. A system in which government ministers transmit orders to their representative, the *préfet* who, in turn, transmits orders further down the chain of command, is a system well suited to the demands of a state bent on military consolidation and agrandisement. It is not, however, flexible enough to respond to rapidly changing economic circumstances nor to satisfy the aspirations of those parts of the country whose regional identity is strong and resistant to suppression. The imposition of this system, and its survival over generation after generation, is not in itself sufficient to explain the emergence of regionalism as a significant factor in the political life of the nation. However, the combination of a rigid administrative hierarchy with uneven economic development which itself stimulates regional particularisms, provides fertile ground for the growth of regional assertiveness.

The structure laid down in 1789-90, consolidated by Napoleon during the Empire, was subject to modification during the course of the 19th century, although not to fundamental change of form. The chief innovations concerned the introduction of the democratic principle. The *département* achieved formal juridical status (*collectivité territoriale*), and the election of its own representative council (*conseil général*) in 1871, while a representative body, the *conseil municipal*, was instituted in the *commune* in 1874. These two laws form the basis for local administration in modern France.

Together with the hierarchical relationships between Ministers, Prefects and Mayors, they encapsulate the complex and sensitive balance between the authority of the state and the individual freedoms enshrined in the principle of election. For reasons which will become clear in the section which follows, the pattern of French local administration remained unchanged from this point until the third quarter of the 20th century. Indeed, given the fact that reforms in this field, whether minor or radical in scope, have always tended to supplement rather than replace existing forms of organisation, it can be argued that even the most far-reaching measures, such as those enacted in the early 1980s, are inevitably less radical than they appear at first sight.

The gradual awareness that the intermediate link between *commune* and state, the *département*, was inadequate to meet the demands of changing economic, demographic, technological circumstances, stimulated at intervals, throughout the 19th century and beyond, a range of propositions for administrative reform from across the political spectrum. Nationalists, monarchists, federalists, and socialists, all had a point of view on the desirability of a new, regional structure. Pierre Joseph Proudhon (1809-65), critical of the centralising nature of the Republic, saw the possibility of new freedoms in a federation of states based on the Provinces of the Ancien Régime. Auguste Comte (1798-1857), whose determinist philosophy emphasised the links between man and *milieu*, foresaw a regionalism which would serve the interests of that determinism. In its practical implications, the monarchism of Maurice Barrès (1862-1923) was not dissimilar, with its insistence on the specific attributes of each region and an almost visceral attachment to diversity and pre-Revolutionary tradition. Less visionary, more acutely politicised, were the polemical views of Charles Maurras (1868-1952), who found freedom incompatible with a democratic republicanism requiring a strong centralised bureaucracy in order to maintain its power. The piecemeal nature of such contributions to a debate on the merits of structural reform of territorial administration explains to some extent the absence of concerted pressures for reform. In the century from 1850 onwards, some 30 private initiatives for regional reform were forthcoming, with another 15 or so emanating from individual parliamentarians (Brongniart, 1971, 46-48). And yet, even with the foundation of pressure groups such as the *Fédération régionaliste française* in 1900 by Charles-Brun, which led to the publication of a detailed manifesto stressing the need for both administrative and economic decentralisation, there is still no evidence of the impetus towards serious reform at governmental level.

The First World War provided some of that impetus; in 1917, the Minister for Commerce, Clémentel, proposed 'economic regions' which were to consist of a regional grouping of local Chambers of Commerce, with an advisory and consultative function. Their influence, like their resources, was limited but represented nonetheless a quasi-official recognition of a territorial division larger than the *département* (Gravier, 1970, p. 47). The massive disruption caused by war was again the occasion for experiment following the division of France into 'Free' and 'Occupied' Zones in 1940, and the establishment of the French State centred on Vichy. In most respects, the changes made to rules governing *communes* and *départements* were retrograde – laws enacted in 1940 and 1942 meant that mayors were henceforth to be appointed, rather than elected, and that the local election of municipal councillors was abolished for *communes* with more than 2,000 inhabitants. Superficially, the creation in 1941 of regions and regional prefects appeared to offer progress in the direction of larger, supra-departmental divisions, but these were intended as nothing more than an extra layer of administration to carry out the wishes of the state. In the aftermath of the Liberation, the regional prefects of the Vichy regime were replaced with regional *commissaires de la République* whose role was to supervise an orderly return to the legality of Republicanism and the subsequent preoccupation with re-establishing of order and central control meant that every trace of regionalisation was under assault. Public order requirements led to the creation in 1948 of administrative regions which coincided with the existing military regions, and the installation at their head of Inspectors General of the Administration in Extraordinary Mission (IGAMEs). While these innovations offered a model for a potential rethinking of the administrative structure of France, they also reflected the lack of will for debate on radical structural change. Experience of the vulnerability of the state, a vulnerability demonstrated in ample measure by the events of 1940-5, by invasion, annexation, occupation, all of which led to fragmentation of the *Hexagone* and the loss of national integrity, ensured the attraction of institutional conservatism. Thus it was, that at the beginning of the Fourth Republic, there were very few voices raised in favour of a major assault on the institutions of the defunct Third Republic. The fear of regional anarchy, the fear of a 'federalism that was such a threat to national unity in times of crisis' (Schmidt, 1990, p. 73), easily prevailed.

Throughout the 19th and until the mid-point of the 20th century, new approaches to the administrative structure of France, and therefore new ways of reconciling the age-old confrontation of national imperative and regional aspiration, remained stuck at the level of theory and debate. It was only in the years following the Second World War that the problems of modernisation, industrial and economic development, and social restructuring, provoking in their turn the resurgence of previously somnolent, or suppressed, regional identities, made themselves felt with such urgency that the prospect of regional reform rose to the top of the political agenda. Two problems in particular came to the fore. An uneven geographical distribution of economic activity, leading to ever more strident contrasts between affluence and poverty on the map of France, led to the overburdening of Paris and its surrounding region, with all the attendant problems of overcrowding, excessive demands on housing, building land, industrial development, transport and communications, and a declining quality of life. We shall now examine in more detail the effects of this uneven economic development, of Parisian domination, and the consequences of the assertion of regional identity.

The Impact of Economic Development

In one sense, the division of France into *départements* was both logical and simple – the fairly even distribution of population in an almost entirely rural country enabled the creation of homogeneous units not widely divergent in size (between 4,500 and 9,000 km^2). The combination of political centralisation and economic growth, however, soon brought about disparities, the effects of which were substantial and lasting. The long period of political isolation during the Napoleonic wars, together with a protectionist trading policy, ensured that the Industrial Revolution got under way later in France than, for example, in Britain. With increased economic and industrial activity came a growing demand for transport. Hence the increase and improvement in existing means of transport, consolidation of the road network and inland waterways, but also the design and construction of the railway network, the famous spider's web of lines radiating out from Paris to the newly flourishing large provincial towns. Modern means of transport – the railways, but also the expanding departmental road network – reinforced in economic and commercial terms the pre-existing political dominance of Paris. Outside Paris and a few large towns, and at least until the early years of the 19th century, France had been predominantly composed of isolated and static rural communities. From then on it became subject to a process of economic and industrial development which altered the relationships between one part of the country and another (Price, 1987, pp. 3-44). Heavy industry was concentrated at first around the major sources of raw materials, in the coalfields of the North, around Lille, in Lorraine with its rich deposits of ore, and along important traditional lines of communication such as the Rhône, around Lyons, with its new textile industry.

Urbanisation was accompanied in France by the concentration of activity in Paris and its surrounding region, and therefore an influx of population, which imposed enormous demands on the capital. Figures relating to the growth of population both nationally and by *département* are instructive in this respect. During the period from 1862 to 1982, the total population of France rose from 37.386 million to 54.338 million (56.6 million in 1990). The increase to 1982 is therefore of the order of 45 per cent. If the growth in the population of the Paris region had reflected this overall pattern, an increase from 2.466 million to 3.584 million could have been expected over that period. The actual figure for 1982 was 9.186 million, an increase of 272 per cent, or more than 5.5 million more than suggested by the figures for the country as a whole. This 'excess' growth was greater than the combined notional surplus in the same period of all the other *départements* with greater than average growth (Lévy, 1982, p. 64). At the other end of the scale, the effects of the rural exodus can be seen in the decrease in the population of *départements* such as the Lot and the Dordogne, sandwiched between the Massif central and the Atlantic coast, which together suffered a 30 per cent drop in population. A stark contrast emerges, too, in a comparison between the most and least populous *départements* outside the Paris region. Confirming the pattern of rural depopulation and migration to the industrial heartlands, the Lozère, at the southern end of the *Massif central*, has a mere 74,000 inhabitants, while the Nord *département* has more than 2.5 million. These two extremes exemplify an imbalance not foreseen by the architects of the departmental system 200 years ago. Furthermore, it is the young and active who migrate from country to city, attracted by the prospect of higher living standards and a working life less restrictive than that to be found on the land. They leave behind an ageing rural population poorly equipped, in every sense, to cope with the demands of modernisation and renewal.

The result is a map of France in which the earlier picture of a stable, agrarian, society is replaced by one showing concentrated areas of intense activity. Paris and the large urbanised industrial centres in the North, the East and, to an extent, the South-

East, exerted a magnetic attraction, and the insatiable demand for industrial and commercial labour began a long process of migration from the rural world to the urban, from small town to large town, from the provinces to the capital. The social dislocation which is the human reality behind the documentary and statistical evidence is chronicled by 19th-century writers such as Balzac, with his *Comédie humaine* novels, laying bare ambition and desire in the pursuit of fortune, and Zola, whose *Rougon-Macquart* cycle portrays the degraded condition of the new urban poor with a fatalistic determinism.

The transformation of France from a rural country to one with a strong urban base is possibly the single most important factor in the forging of a cohesive nation (Weber, 1976). The process of urbanisation which gathered pace in the last decades of the 19th century coincided with the political decisions on educational uniformity and the inculcation of the republican ideal which was the hallmark of the early years of the Third Republic. It would be wrong to assume, however, that the transformation happened overnight, or that it can be described as a simple transfer of population from country to city. Well into the 20th century the image of France as a nation of peasants remains convincing – in 1914 the ringing of church bells summoned peasants from their fields to the village and enlistment in the army. The process of industrial development itself did not imply instant urbanisation, since much early industrial activity was of an artisanal nature in sectors such as textile production. This meant that some industrial activity took place in a rural environment, with work moving in the direction of the labour force, although the countervailing trend of movement off the land towards the towns became the dominant feature, and many migrants came to the towns with some previous experience of artisanal or light industrial work (Merriman, 1982, p. 23).

The phenomenon known as *l'exode rural* (rural exodus) goes hand-in-hand with the progress of the industrial revolution. At the beginning of the 18th century, France had a population of around twenty million, of whom seventeen million could be classed as rural (i.e. living in *communes* of fewer than 2,000 inhabitants), and fifteen million agricultural (i.e. living from farming). While this population increased numerically up to the middle of the 19th century, its size as a proportion of the total population fell, given the faster rise in the number of town dwellers. The agricultural population fell from 70 per cent of the total in 1789 to around 55 per cent in 1850. It was the poorest regions with the least productive, mountainous, terrain which provided the new urban and industrial workforce. The reduction in the numbers working in agriculture was a constant feature from the 1850s – by 1946 the figure was little more than 10 million, and this represented just over half the still substantial rural population, and a quarter of the total population of 40.5 million. Thirty years later this figure had fallen much further, with 5,884,000, or 11.5 per cent of the population, dependent on farming, and no more than 9 per cent actively engaged in agriculture (by 1990 that last figure had fallen to 5.6 per cent). By contrast, the urban population, which two centuries ago accounted for less than 18 per cent of the total, by the 1980s accounted for about three quarters, an increase which, while substantial, still leaves France less urbanised than many other European countries.

The difficulties created by the unbalanced industrial and economic development of

France were not addressed until after the Second World War. This inactivity can be attibuted to several factors which, while discrete, contribute to the creation of a form of myth or idealisation. In the wake of the traumas of the Franco-Prussian war of 1870, and of the war of 1914-18, there was some reassurance in the image of a France basking in its geographical and climatic advantages and taking comfort in the traditions of rural life. (The substitution by the Vichy regime of *Travail, Famille, Patrie* for the republican motto *Liberté, Egalité, Fraternité* bears witness to the emotive appeal of deeply-felt traditional values in times of crisis and uncertainty.) The insistence on France as the epitome of equilibrium and harmony, combined with elementary ignorance of the realities of agricultural inefficiency and low productivity, allowed a certain complacency to persist. The more or less even balance between the urban and the rural (the urban population only began to exceed the rural from the 1930s) was seen as a positive strength, while subsistence farming and erratic polyculture were confused with efficiency. Right up to the middle of the 20th century and beyond, inadequate or non-existent economic planning reinforced the tendencies of the previous century – most modern industrial activity (cars, aeronautics, chemicals) was centred around Paris. The massive natural resources in the North and the North-East were exploited less effectively than they might have been, due to the strategic hesitations induced by the proximity of vulnerable borders. It has become something of a cliché to refer to the division of France into two parts, one prosperous, the other under-developed, by a diagonal line drawn roughly between Le Havre and Marseilles. During the Third Republic, however, and for much of the Fourth, there was little to counter the relentless accentuation of the contrast between an active and prosperous France in the North and East (including, of course, Paris), and a virtual desert in the South and, in particular, the West where, with the exception of a few large ports, industrial activity was non-existent. In the years following the Second World War, the conservative instinct which had rejected administrative and institutional change in favour of structures copied from the Third Republic was at work in the minds of those who had the responsibility for the process of economic renewal. The innovative opportunities offered by post-war reconstruction were, on the whole, not taken – the objective was to reconstruct on the pre-war pattern (Drevet, 1988, p. 25).

There is however one factor which, taken together with an accelerating speed of technological change, a new Industrial Revolution, finally forced governments of the Fourth and Fifth Republics to embrace concerted economic planning strategies and the need for a rational organisation of French territory. The population of France had remained static for the first half of the century (40.7 million in 1901; 40.5 million in 1945). In the 30 years that followed, the increase was dramatic (to 52.65 million in 1975) and made change inevitable, particularly since the urban population rose twice as fast as the rural. This 30 year period – *les Trente Glorieuses* – saw the transformation of French society, with dramatic increases in living standards, productivity and efficiency (Fourastié, 1979). A rural society shifted its focus to the urban world, and began the process of adaptation and modernisation which would make necessary a re-evaluation of the unequal relationship between Paris and *la province*. The publication in 1947 of *Paris et le désert français* by Jean-François Gravier marks a watershed in this process, capturing the imagination of political élites and underlining both the uneven and inadequate development of provincial France and the harmful and debilitating effect of Paris on the rest of the country. A capital city, *la ville lumière*, which for generations, indeed centuries, had been seen as the glory of France, the seat of industry, dynamism and wealth, the guarantor of national security, now found itself accused of bleeding dry the provinces, of sucking in the resources and talent from all corners of the *Hexagone*. Gravier far-sightedly perceived the need to mount an assault on an ingrained provincial inferiority complex by holding Paris, as seat of government and centre of administration, responsible for the excesses of one centralising regime after another. The problems created were economic, political and administrative, and responses addressed all three factors as, in the course of the Fourth and Fifth Republics, governments sought to plan for rational economic development and to come to terms with a range of regional aspirations. An assessment of these responses forms the substance of the next section.

Regional Reform: the Early Stages

Three factors in particular ensured that, during the course of the *trente glorieuses*, the problems of regional imbalances and of the dichotomy between Paris and *la province* would have to be addressed. First, the post-war surge in the birth-rate, *le baby boom*, continued far beyond what might have been expected in the euphoria of the Liberation. For 28 consecutive years, from 1946 to 1973, the annual number of births exceeded 800,000 (the pre-war figure was around 600,000). Increasing prosperity meant that mortality rates declined in the same period and, although the birth-rate pattern was not to be repeated in the following generation, in spite of a natalist policy offering substantial financial incentives to families, other factors, such as a policy of substantial immigration from Algeria, Spain and Portugal, as well as the impact of returning French Algerians following that country's independence in 1963, led to a continuing rise in the population (Lévy, 1982, p. 13). The effects were, as always, felt disproportionately in Paris and the surrounding region – predictions of 14-16 million inhabitants by the end of the century were taken seriously. But pressure was also put on administrative structures throughout France, ensuring that the *départements*, which previously had performed the role of simple administrative relay between state and citizens, had to confront the demands for services in the broadest sense. Economic development aid, infrastructure projects, investment in housing, schools, and public facilities of all kinds brought to the surface the need for concerted planning and a rational distribution of resources.

The second factor relates to the widening of horizons that comes with industrial and technological innovation; the France of the Fourth Republic was a France in which people in large numbers became mobile, travelling to work, or for pleasure at the weekend, travelling the length of the country on their annual holiday, embarking on the N7 (*Route nationale 7*) from Paris to the South, in their 4CV, Dauphine, Simca Aronde, or DS. The first-hand knowledge of somewhere other than the home *quartier* developed a public awareness of difference, and the realisation that the map of France had a certain elasticity – good communications shortened distances and improved efficiency. Poor communications had the opposite effect, and certain parts of the country, poorly served over the years by inadequate investment in roads, railways, and the rest, lagged behind. Large numbers of rural *communes* were ill-equipped, or too small, to deal with complex problems of resource management. This gave rise to the process of *fusion* of small *communes*, often in the teeth of much local rivalry, so that the total number has been gradually reduced to around 36,000. In spite of this rationalisation, only 2 per cent of *communes* have more than 10,000 inhabitants, and 90 percent still have fewer than 2,000, while nearly a quarter of the total, nearly 10,000 *communes*, have fewer than 200 (Gruber, 1986, p. 180). Former Interior Minister Pierre Joxe tells of the problems he encountered in trying to propose a radical reduction in the number of *communes* through intercommunal cooperation. Despite his failure to convince President Mitterrand or colleagues in the post-1981 government such as Gaston Defferre or Pierre Mauroy, all of whom were wedded to the existing communal system, he remains convinced that, sooner or later, 'France will have 4,000 or 5,000 *communes* instead of the present 36,000. The only question is which millenium it will happen in. It's too late for the second, but a possibility for the third. Or fourth' (*Le Monde*, 26 May1998, my translation).

Third, and in apparent contradistinction to the previous point, the expression of regional identity took on a vigour which had not been very visible during the century and a half of republican uniformity. The desire on the part of individuals and groups to proclaim and to take pride in local and regional values and traditions gained a ground-swell of support, and not only in the more peripheral regions, such as Brittany, Alsace, Provence, the Basque country or Corsica, but also, paradoxically, in the capital. A capital, 60 per cent of whose population was born in the provinces, was likely to harbour among its inhabitants a substantial nostalgia for regional variety and independence of spirit, coupled with a strong if unrealistic desire for an eventual *retour au pays* (return home).

It is unsurprising, then, that the first initiatives for regional reform came piecemeal, in the shape of private ventures – the grouping together of chambers of commerce to coordinate, in embryonic fashion, efforts for improved regional development. Foremost among these was the *Comité d'Étude et de Liaison des Intérêts bretons* (CELIB) which, under the chairmanship of René Pleven, produced a programme for regional economic action in 1952. This preceded by two years the official recognition of *Comités d'expansion économique*, formed by a heterogeneous mixture of mayors, local councillors, industrialists and members of chambers of commerce with the aim of acting as regional pressure groups. By this time, the need for some form of economic planning was well entrenched at national level, but the first of what was to become the series of national economic development plans (the *Plan Monnet* of 1946-7) contained no element of regional discrimination. As Gravier has pointed out, the phrase 'town and country planning' was in use in Britain well before the 1939-45 war, but the notion of *aménagement du territoire*, which carries much of the same meaning, and adds the concept of local economic development, only appeared in France towards the end of the war, and had little impact then or for some time afterwards (Gravier, 1970, p. 57).

In 1955, the pressure exerted by the *comités d'expansion économique*, combined with the need for the state to find an appropriate framework for its economic management, led to the setting up of 22 *circonscriptions d'action régionale*, after much debate about the appropriate size and number of the new territorial divisions. These have remained the basis of regional organisation, despite revision and modification (the original Rhône and Alpes regions were brought together to form Rhône-Alpes in 1960, and Corsica was separated from Provence-Côte d'Azur in 1970). In 1964 a *Préfet coordonnateur* was installed in each region, fitting neatly into the hierarchical control system enshrined in the existing relationship between state, *département* and *commune*. At the same time, each region was provided with a *Commission de Développement économique régional* (CODER), made up of a mixture of elected representatives and interested parties nominated by the government, but with no budgetary control, and only a consultative role. The contrast between these initial regional institutions and the *Délégation à l'Aménagement du Territoire et à l'Action régionale* (DATAR), set up as an adjunct to the Prime Minister's office in 1963, is instructive, and defines the contrast between two forms of potential action for regional development. The DATAR, acting through the various ministries and regional and departmental prefects, ensures that decisions taken centrally about investment priorities are transmitted directly to the relevant sector, a process of *déconcentration*. A system which permitted, or encouraged participation in that decision-making process at local level, and allowed finance to be raised and resources to be used at the discretion of directly elected local bodies, would qualify as *décentralisation*. All the early moves towards decision-making in regional development are categorised by the former rather than the latter term.

It would be an exaggeration to claim that the social and political upheaval that convulsed France in the spring of 1968, *les événements de mai*, had its roots in the question of regional self-assertion and the frustrations brought about by economic and cultural impotence. Nevertheless, the image of a state frozen in the rigidity of its structures, with its élites unresponsive to the aspirations of many of its citizens, is very much that of France at this period. It is significant, therefore, that a speech made by President de Gaulle in March 1968, in Lyons, offers a recognition of the changes called for in terms of territorial organisation:

> The centralisation which for centuries was needed for our country to achieve and to maintain its unity in spite of the diverse nature of the provinces which, one after another, had been added to it, is no longer essential. On the contrary, regional activity now appears as the springboard for its future economic success (Brongniart, 1971, p. 3, my translation).

The following year, de Gaulle's plans for regional reform (combined with a putative reform of the *Sénat*), were subject to a referendum. The plans contained novel features – the entitlement to receive the proceeds of one or two (unspecified) state taxes; the power to borrow money; full status for the regions as *collectivités territoriales* on a par with *départements* and *communes* but, unlike these latter and in apparent contradiction of the Constitution of 1958, with no elective element. In all, the intention seems to have been to encourage greater regional participation in decision-making. In the event, the pre-referendum debate was reduced, in part at least, to a sort of plebiscite on the future of de Gaulle himself, with the result that the rejection, by 53 per cent of the voters, was perhaps less a comment on the merits of regional reform than a judgement on the immediate political situation. The measures initiated under de Gaulle's successor, Georges Pompidou, were in fact less far-reaching than those rejected in 1969. The Region was to be seen as 'the concerted expression of the *départements* of which it is composed' (Dayries & Dayries, 1978, p. 40), rather than an additional layer of administration with a status of its own. Nevertheless, a change of mood is evident from the early 1970s. Burgeoning enthusiasm for the expression of regional identity and culture is no longer stifled. Regional languages start to appear as options on programmes of study for schools and universities.

Throughout this period, however, it is difficult to discern any clear statements of policy on the regions from the main political groupings. If anything, there is a collective, if unformulated, reiteration of the attitudes outlined by de Tocqueville 150 years previously – those with their hands on the levers of centralised power remain antipathetic to the theme of regionalism, as it represents a threat to that power. Throughout the 1970s the inheritors of the Gaullist tradition (the *Union des Démocrates pour la République (UDR)*, which became the *Rassemblement pour la République (RPR)* in 1976), remained in favour of centralisation, a policy which, with its roots in the re-establishment of republican unity after the war, was unsurprising. The *Parti républicain*, under the leadership of Giscard d'Estaing, was similarly unimpressed, at least while in office as part of the parliamentary majority. Among this majority, only the small centrist grouping, given voice by Servan-Schreiber and Lecanuet, was open to the discussion of regionalist themes. On the left, too, regional matters gained little attention. The *Parti communiste*, centralist to the core, was in any case preoccupied with effecting a change of government at national level, and the parties of the left in general shared that preoccupation, holding at arm's length those regionalist movements with which they might have been expected to make common cause. In addition, the structure of all the main national political parties reflected the centralised structure of the state so that they were ill-fitted for regional modes of operation. This is not to say that no thought was given to the theoretical possibilities of institutional change. For many individual political figures on the left (e.g. Gaston Defferre, Michel Rocard) the need was for fundamental change to shake off the Gaullist tradition (Mény, 1974, p. 394). The publication of *Les Citoyens au Pouvoir* (Club Jean Moulin, 1968) crystallised the argument in favour of a massive reduction in the number of *communes* and the creation of a reduced number of powerful regions (the subtitle was *12 Régions, 2000 Communes*). In time, however, through the 1970s, opposition parties as a whole began to see the regionalist theme not simply as a distraction from the priorities of national politics but, given the persistence in government for two decades of a right and centre majority, as a possible weapon for an assault on this majority, a means of bringing about change at a national level. This perception, together with the growing tendency for regionalist movements to create an impact on the national stage, by spectacular protest and in some cases violent outburst, contrived to bring regional reform to the fore.

Regionalist Movements

Regional rebellion took a number of forms during the 1970s. As with the impetus to regional economic reform (the CELIB), it was Brittany which led the way, building on a traditionally uncompromising rejection of the authority of the French state, proclaiming a Breton nationalism which, as with many other regionalist/nationalist movements, had its origins on the political Right and its modern manifestation on the Left. At municipal elections in 1977, the *Union Démocratique Bretonne* enjoyed a measure of popular success, having 33 of its candidates elected. This popular support, stimulated by the crisis in Breton agriculture in the 1960s and 1970s, but drawing also on an appeal to the assertion of cultural and linguistic difference, was echoed in only one other French region, Corsica, where similar agricultural problems provoked a similar response. In Brittany as in Corsica, regional activism did not restrict itself to the legitimacy of the ballot box. The *Front de Libération de la Bretagne* made a number of small-scale but noisy attacks on symbols of French centralist power from the late 1960s onwards (television transmitters, tax offices and the like). The immediate cause of discontent leading to the expression of regionalist/nationalist sentiment in Corsica was not so much the familiar effects of centralisation, the *exode rural*, the ageing population, under-investment, indiscriminate application of national policies, as the specific instance of the repatriation of thousands of French from North Africa and their virtual takeover of the Corsican vineyards. The *Société pour la Mise en Valeur de la Corse* (SOMIVAC), established to provide centralised support for modernisation of the island, effectively subsidised the purchase of land by the new arrivals, and alienated the existing population. A welter of legitimate political groupings for the defence of Corsican interests, notably the *Action pour la Renaissance de la Corse* (ARC) received considerable support, but outbreaks of violence marked the division between those in favour of gradual development within the law and the outright autonomists. The result was twofold, and this is symptomatic of the situation in Brittany, although there its effect was felt to a lesser extent. On the one hand, the government clamped down hard on extremist action, while on the other it undertook a number of measures (job creation, increased equipment budgets, establishment of a university at Corte) designed to placate popular opinion and deprive the activists of support. The continuing if sporadic eruption of violent incidents, such as the assassination of the island's *Préfet* in February 1998, suggest that the policy has been less than entirely successful, while the arrival in Corsica of both President Chirac and Prime Minister Jospin in the immediate aftermath of the attack reflects a clear institutional commitment to republican unity.

In addition to the Breton and Corsican regionalist movements, mention should be made of Occitan regionalism, which also displays a movement from its roots in the nationalist Right towards the socialist Left but which, to a large extent, was confined to the essentially cultural and linguistic preoccupations of an intelligentsia, and only attained a level of serious popular support with the crisis in viticulture in the late 1960s and 1970s. Its principal apologist was Robert Lafont, whose *La Révolution régionaliste* (1967) and *Décoloniser en France* (1971) provided the polemical base for the autonomist movement under the banner *Volem viure al païs* and for his own short-lived political ambitions at the national level.

Regionalist fervour in Alsace has been relatively muted; indeed, the history of the region, with its alternation between French and German sovereignty over the generations, has meant that regional instincts are rather favourable towards continued integration with the French nation. Nevertheless, linguistic and cultural distinctiveness was channelled into pragmatic concerns about employment and, more recently, ecological matters, while not hindering cross-border cooperation with neighbouring regions in Germany and Switzerland.

In Catalonia and the Basque country, the existence of regionalist sentiment was clearly less significant in the French context that it was for Spain. The Basque population of South-West France is only 170,000, compared with 2.5 million on the Spanish side of the Pyrenees. For both the Basques and Catalans in France, the question was one of identity, self-assertion and cultural survival, manifested by the existence of regionalist pressure groups and, from time to time, their translation into political movements of the Left which tended to receive only minimal electoral support.

Economic Crisis and the Pressure for Reform

The increased activity of, and support for, regionalist pressure groups during the 1960s and 1970s is accompanied and no doubt influenced by profound economic changes. France's founder membership of the then European Economic Community had already started to exert pressure for reform and modernisation on agriculture, notwithstanding the protection afforded in the early years by the Common Agricultural Policy. Projects for restructuring small and inefficient farms (*remembrement*), particularly in the south and west, were more and more numerous and were seen as a way of coping with an ageing farming population (in 1980, 40 per cent of male farmers were in the age range 55-75). Many of the traditional bases of French industrial activity, such as coal and steel production, and textiles, were under pressure from foreign imports, and France's dependence on imported energy sources and raw materials ensured that, while industrial expansion continued into the 1970s, the years of recession which followed the oil crises of 1974 and 1979 exacted a heavy toll in terms of output. The annual percentage increase in GDP, which had averaged 4.2 per cent during the 1950s, peaked at 5.7 per cent in the 60s, before falling to 3 per cent between 1973 and 1979, and to 1.5% after 1980. Throughout this period, the growth in the tertiary, or service sector tended to mitigate the effects of the traditional north-east/south-west economic divide, bringing increased economic activity to towns all over the France. Particularly well placed to benefit were those towns selected in the mid-1960s as a focus for government investment in an attempt to counterbalance the overwhelming economic presence of Paris. Lille, Nancy and Metz, Strasbourg, Lyons, Marseilles, Toulouse, Bordeaux, Nantes and St Nazaire all enjoyed for a time the status of *Métropoles d'équilibre*, a policy which both asserted a latent provincial vitality and, perhaps inevitably, ensured that distinguishing features of the economic life of those towns gradually disappeared under a uniform blanket of nationally and multinationally-inspired investment.

As for Paris, its long-established role as a magnet for internal migrants appeared to be coming slowly to an end. An inexorable rise in population over 150 years which, by the early 1980s saw 18.6 per cent of the population (10.056 million) concentrated on a mere 2.2 per cent of the territory, was halted. In the seven years between 1975 and 1982, the Ile de France, as the Paris region was known from 1972 onwards, having had the highest rate of net immigration, became the region with the highest rate of net emigration, shrinking by 435,000. This represented a net loss of 4.4 per cent, compared with, for example, a net gain over the same period of 6.35 per cent in Languedoc-Roussillon. Over a 20 year period up to the mid 1980s, Paris lost about a quarter of its industrial manpower (Gravier, in Uhrich, 1987, p. 325). The capital's dominance in the tertiary sector, however, remained massive – in banking, insurance, and the all-important head offices of major national and multinational companies, as well as in research and higher education. Clearly, the economic map of France had been undergoing major change and the distinctions previously visible were giving way to more piecemeal and complicated divisions. Stagnation in the North and East (Nord, Champagne-Ardenne, Lorraine) following the downturn in heavy industry and the tendency of the young to leave in search of work; stagnation because of an ageing population and low birth-rate in much of the *Massif central* and parts of the South-West (Auvergne, Limousin, Midi-Pyrénées); large population increase along the Mediterranean littoral (Languedoc-Roussillon, Provence-Alpes-Côte d'Azur); net migration into the regions of the Atlantic coast (Bretagne, Pays de Loire, Poitou-Charentes, Aquitaine); and continuing dynamism in the Centre and Rhône-Alpes.

It is against the background of these changes that François Mitterrand opted to make the need for thorough regional reform a major feature of his programme for the presidential campaign of 1981. '*La grande affaire du septennat*' was the phrase used to describe its centrality to his first term of office. The legislation that followed Mitterrand's victory and the subsequent victory of the Left in the legislative elections was many-sided and complex. Its impact overall was to introduce a certain coherence into French local administration, putting *communes*, *départements* and *régions* on an equal footing as *collectivités territoriales* and making clear their specific attributions and functions. The *commune* assumed responsibilty for *urbanisme* (urban planning and building applications), for elementary schools, local roads, water, sanitation and drainage services, and urban transport. The *département* dealt with a wide range of social services (child support, health care, homes for the old and handicapped), for 11-15 education (*collèges*), for departmental roads and non-urban transport. The *région* had responsibility for post-16 secondary education (*lycées*), training, regional economic development (*aménagement du territoire*), regional transport, and the establishment of periodic planning contracts (*contrats de plan*) between region and state. Within this framework, responsibility for the universities, the high-speed train network (*TGV*), main roads (*routes nationales*) and motorways lies with the state. Underpinning this breakdown of responsibilities were two innovations which defined the radical nature of the reforms. For the first time, power was devolved to directly elected regional assemblies, and the role of the *préfet*, at both regional and departmental level was substantially curtailed. The *préfet* was no longer to be the all-powerful representative of central government, but an adviser and observer, whose authority was limited to an *a posteriori* control of the legality of decisions taken by councillors or mayors. As a symbol of this reduced role, the name *préfet* was jettisoned in favour of the cumbersome *commissaire de la République,* a curious echo of post-war arrangements and one which the right-wing government of Chirac, during its period of *cohabitation* with Mitterrand from 1986 chose to reverse. It did not, however, make any attempt to reverse the nature of the reforms, which seem to have been widely accepted as a sensible modernisation of a stiff and archaic structure. Additional elements of the reforms were significant in that they created a breach, for the first time, in the monolithic and indivisible republic. Corsica received a special statute (*statut particulier*), with provision for generous *per capita* representation on its regional assembly and for elections scheduled several years earlier than those for the other regions. Similarly, a special statute for Paris, Lyons and Marseilles allowed for a mini-decentralisation within those three largest cities, with directly elected district councils (*conseils d'arrondissement*). Legislation to restrict the traditional *cumul des mandats*, by which individuals could hold elected posts at a variety of local and national levels simultaneously, was enacted, to the surprise of no-one, with a certain lethargy. Pre-1981 hints by Mitterrand that a *département du pays basque* might be forthcoming were not developed.

A number of observations suggest themselves in the wake of these reforms. The euphoric assumption, in the early 1980s, that they would herald a liberation of previously untapped sources of dynamism and involvement has been dampened by the undeniable reality of a macro-economic climate that has relegated the entire regional question to the margins of debate. The regional elections which took place in 1986 (confusingly in the shadow of the parliamentary elections) and again in 1992 and 1998 saw neither the emergence of a new political class, nor specifically regional voting trends – people voted, on the whole, in much the same way as they would have voted in national elections (Perrineau, 1987). However, the consequences of a proportional voting system which, particularly in 1998, allowed the extreme right *Front national* to gain influence and sow confusion among the other major parties of the right willing, in some regional assemblies, to accept its support in order to achieve a majority, have yet to become clear. On the other hand, the various shifts in the national political mood reflected in the parliamentary elections of 1993 and 1997 revealed no obvious desire to dismantle the reforms of the 1980s. Indeed, following the victory of the Right in 1993 the *Ministre de l'Intérieur et de l'Aménagement du territoire* in the government of Edouard Balladur, Charles Pasqua, was at pains to reclaim the regional development standard for the Right, while starting to sketch policies which take account of a France whose future economic strengths lie not simply in the big urban concentrations but in innumerable smaller networks of activity. The intention appeared thus to be the 'reconquest' of the territory as a whole, with each part brought within reach, not of the capital or the big cities, but of provincial centres of communication and resources (*Le Figaro*, 1 July 1993, p. 6). Little remains of these initiatives, however, following the election of the socialist government of Lionel Jospin, although the traditional fault line which separates the proponents of *décentralisation* and *déconcentration* remains clearly visible. The new Interior Minister, Jean-Pierre Chevènement, a vigorous defender of the nation state, has appeared reluctant to jettison the concept of a national plan for regional development enshrined in the Pasqua legislation. Nonetheless, a fresh perspective on *l'aménagement du territoire* can be discerned in plans put forward by Dominique Voynet, the Minister whose brief includes both *aménagement* and the environment. These envisage the creation of 22 separate regional development plans in place of the traditional single national plan. Most significantly, there is to be a new emphasis on the role of towns in sponsoring and motivating regional development and regeneration. This is proposed as a positive alternative to earlier policies (seen as 'ruralist') which tended to concentrate on alleviating rural underdevelopment and compensating deprived areas. Earlier regional development policies are seen, also, as having concentrated too heavily on costly infrastructure projects (derided as 'concrete and papering over the cracks') and the intention now seems to be to give priority to more basic provision of services (*Le Monde*, 19 February 1998 and 19/20 April 1998).

From one perspective, this approach can perhaps be seen as a response to the objection which has often figured in discussion of regional development, that *décentralisation* and *aménagement du territoire* are fundamentally incompatible, because the first implies autonomy of decision at the periphery, while the second requires decisions to be made at the centre. However, perhaps the most interesting implication of the continuing vitality of the topic is that, wherever the decisions are made, it would seem that there is no longer the automatic assumption in the regions that Paris is to be the functional focus of activity. This perception may, arguably, be flying in the face of economic reality since, once again, as the French economy picked up in the second half of the 1980s, it was Paris that was the first to benefit – nearly half the jobs created in the years after 1986 have been in the single *département* of Hauts de Seine (Benko & Lipietz, 1992, p. 13). There can be little doubt, however, that regional development is judged by government in the late 1990s primarily on its capacity to create employment. And yet in all sorts of ways that transcend the immediate economic situation, the regions of France have been shifting their gaze from the capital to their neighbours and, in some cases, the neighbours are not exclusively French.

The Interregional and Cross-National Dimension

For more than thirty years France has pursued policies intended to reduce economic disparities between regions. For most of that period, the policies have been prepared and monitored by the DATAR in Paris, exercising its judgement about the relative merits of programmes of action affecting different parts of the country. At the same time France has had access to a variety of funds set up by the European Community. The Treaty of Rome made no mention of regional preoccupations, and the Single European Act of 1986 was only slightly less laconic, referring to the intention to 'reduce the gap between different regions and the backwardness of the least developed regions'. The Maastricht Treaty had a little more to say on the subject, setting a high priority on economic and social cohesion in the regions of the Community, in order to lessen differences in development and to enable them to profit from the single market and, subsequently, a single currency. Since the inception of the Community, France has benefited from the *Fonds Européen d'Orientation et de Garantie Agricole* (FEOGA-O), for the modernisation of agriculture, and also from the *Fonds Européen Social* (FSE). From

1974 onwards, the creation of the *Fonds Européen de Développement Régional* (FEDER) has also provided resources for specific purposes, in particular, major infrastructure projects and industrial conversion projects (in the steel and textile industry, shipbuilding and fishing). Those funds destined for the Community's poorest regions have by and large passed France by, since none of its regions, with the exception of Corsica and its overseas *départements* (DOM) is poor enough to qualify. However, from 1984, the Community has participated in the co-financing of the State-Region planning contracts (*contrats de plan*), on the basis of priorities fixed by the regions. The amount of money available for infrastructure projects in underdeveloped regions such as the Massif central, and for industrial conversion projects amounts to a sum of around two billion francs a year, which is nearly the equivalent of the entire DATAR budget. In all, these projects affect nearly half of French territory (35.4 per cent is classed as rural development zone, and 12.5 per cent as industrial conversion zone). From 1986 onwards, the *Programmes Intégrés Méditerranéens* (PIM), initiated with the enlargement of the European Community to include Spain and Portugal, have made funds available to the five southernmost regions of France (Aquitaine, Midi-Pyrénées, Languedoc-Roussillon, Provence-Alpes-Côte d'Azur, and Corsica) as well as to the *départements* of Drôme and Ardèche. In the period 1989-93, the total sum made available by the Community to France for structural purposes was of the order of 40 billion francs, while FEOGA funds under the Common Agricultural Policy amounted to about 37 billion francs a year over the same period. It should be noted, however, that access to the various European Community regional intervention funds was, from their inception, channelled through and controlled by Paris. Direct contact between individual regions and the European Commission in Brussels has long been frowned upon, indeed prohibited, by central government – ministerial pronouncements on the subject, even from the government which enacted the reforms of 1982, have been unambiguous. By 1985, however, a less rigid attitude was in evidence, and an inter-ministerial committee for *l'aménagement du territoire* declared that 'the State is no longer opposed to direct contacts between regions and European institutions [...] since no French region has centrifugal desires' (Uhrich, 1987, p. 310, my translation).

As part of the opening out which is implicit in the growing relationships between region, state, and Community, there have been numerous initiatives in France in recent years to develop the regional idea beyond the confines of single regions, and indeed beyond the frontiers of France. In most cases, these initiatives exist at the level of consultation, declarations of goodwill and intent and as expressions of inter-regional solidarity. Among the earliest was the creation, in 1979, of a ten-year plan for the economic development of three border regions, Aquitaine, Midi-Pyrénées and Languedoc-Roussillon, in order to prepare them for the impact of the entry of Spain and Portugal into the European Community. The themes covered by the plan ranged from development of the road system to the promotion of regional wines, from job creation to the encouragement of research potential. With the change of government in 1981, the plan was suspended, but by 1986 the presidents of Aquitaine, Midi-Pyrénées, Languedoc-Roussillon, Provence-Alpes-Côtes d'Azur and Corsica had laid the foundations for *Le Grand Sud*, with its own office in Brussels. This grouping of the five southernmost regions is intended to supply a forum for the discussion of common interests such as transport infrastructure, tourism, training programmes, scientific and technical cooperation, as well as coordinating the approaches of individual regions to Brussels. Another initiative that goes beyond the borders of France is the *Communauté de Travail des Pyrénées*, set up in 1983, which aims to minimise the impact of the Pyrenees as a barrier between regions on either side (Aquitaine, Aragon, Catalonia, Euskadi, Languedoc-Roussillon, Midi-Pyrénées, Navarre, as well as the Principality of Andorra), encouraging the development of those regions, and contributing to the process of European unification. In comparable, if rather less ambitious vein, the *Communauté de Travail du Jura* has, since 1985, dealt with a range of industrial, agricultural, tourism and communications matters common to Franche-Comté and the Swiss *cantons* of Berne, Vaud, Neuchâtel and Jura. The same format has embraced adjacent regions from France, Switzerland and Italy since 1984, with Rhône-Alpes, Provence-Alpes-Côte d'Azur joining the Swiss *cantons* of Vaud, Valais, and Geneva, and the Italian regions of Piedmont, Liguria and Val d'Aoste in the *Communauté de Travail des Alpes Occidentales*. In 1991, a cooperation agreement brought together into a so-called 'Euroregion' five regions from three countries – Nord-Pas de Calais, Kent, Brussels-Capital, Wallonia and Flanders. The intention in this instance is to maximise benefit from the opportunities provided by the Single Market and to capitalise on the physical links provided by the Channel Tunnel and the North European High Speed Train projects. Perhaps the most ambitious of the inter-regional and cross-frontier groupings is that comprising the *Arc Atlantique* which, since its inauguration in 1989, has brought together the 26 regions of Europe's Atlantic coastline, from Scotland to the Algarve. The impetus for this initiative has its origins in the Pays de la Loire region, whose president, Olivier Guichard, is a long time proponent of regional development. The Pays de la Loire had already put in place a number of bilateral trans-national agreements, with Galicia and Andalucia in Spain, Emilia-Romagna in Italy and Schleswig-Holstein in Germany. The initial concerns of the *Arc Atlantique* were focused on the essentially maritime nature of activities common to all the regions involved. The development of ports, the exploitation of marine resources, and environmental protection all figured highly on the agenda. Research, training, and transport infrastructure are seen as the priorities to be addressed in the immediate future.

Inter-regional cooperation was in evidence as early as 1984, when the *Association des Régions Françaises du Grand Est* brought together Alsace, Bourgogne, Champagne-Ardenne, Franche-Comté and Lorraine, a grouping based, like that of *Le Grand Sud*, on the decentralisation laws of 1982 and 1983 which provided for agreements between two or more regions for their common benefit. Research, technology transfer, higher education and training, tourism, and communications form the bulk of the priorities of *Le Grand Est*. Here too, an office has been set up in Brussels. Associations such as *Le Grand Sud* and *Le Grand Est* appear to represent, in part at least, the desire to counter the preponderance of Paris. Ironically, such concerns are to be felt not only at the extremities of France, but in the capital's own hinterland. Since 1990, the eight regions which make up the *Grand Bassin Parisien* (Ile de France, Picardie, Haute-Normandie, Centre, Basse Normandie, Pays de la Loire, Champagne-Ardenne, and Bourgogne) have been in consultation to ensure that the continued growth and prosperity of the first among them (Ile de France) does not work to the detriment of the others (*Le Monde*, 11-12 July 1993, p. 18). Concerted planning for infrastructure, communications, housing, environmental protection, higher education, research, are the familiar themes. In all, 17 French regions have established offices in Brussels.

With the exception of the *Grand Bassin Parisien*, these various schemes, national and trans-national alike, share one striking feature, which has two facets. They concern regions traditionally thought of as peripheral. And they all ignore the capital so as better to pursue dialogue with their equally marginal neighbours, and maintain direct liaisons with the institutions of the European Community in Brussels. In addition, their programmes stress the importance of communications as the key to development. The preoccupation with *désenclavement* (opening out) of previously isolated regions, whether through road, rail, air, or developments in telecommunications, is a constant feature of regional development planning at every level. In 1971 there was no motorway at all to the west of the Le Havre-Marseilles dividing line. By 1983 the DATAR and the Ministry of Transport had produced a *schéma directeur* for the construction of motorways and other high density roads serving France as a whole, and by the late 1980s there were around 6,000 kilometres of motorway, with plans for a further 1,700 kilometres by the year 2000 (Uhrich, 1987, 275-76). Perhaps the most significant feature of these developments is the reduced emphasis on Paris as the hub of the transport system. Admittedly, the High Speed Train (TGV) network radiates from Paris and can be said to confer enormous advantages on the regions that it serves, at the expense of the less prestigious nationwide rail network as a whole. The prospect of through services which permit North-South transit without the need to change trains in Paris is, however, a major feature of TGV plans. More dramatic is the evolution of the road network to a stage which goes significantly beyond the 1983 *schéma directeur*. The construction of motorway links which ignore the Paris dimension entirely – Bordeaux-Lyons via the Auvergne, Mediterranean littoral connections bringing the regions of *Le Grand Sud* within easy reach of Barcelona and Milan – all serve to displace the centre of gravity of French activity and to galvanise interregional and Paris-*province* relations.

Conclusion

It has long been established that the geographical centre of France lies in the Cher *département*, 40 or so kilometres south of Bourges on the RN144. In recent years, a number of *communes* located a little further South and East, in the Puy de Dôme below Vichy, have indulged in friendly rivalry for the trivial distinction of being the geographical centre of the European Community. Even with the opening up of former East Germany and the accession of Greece, the central point has apparently moved only a short distance further east. More generally, the number of towns making the somewhat vacuous claim to be the *carrefour de l'Europe* grows constantly. And Prime Minister Jospin is determined to underline the strategic centrality of France to the European project since, like Europe itself, and alone among Member States, it is simultaneously 'Mediterranean, continental and Atlantic' (*Le Monde*, 18 December

1997). The significant fact to emerge from this welter of enthusiasms is that the centre of gravity of French life, for so long the unchallenged prerogative of Paris, is now less easily determined. The need for a strong, central, unifying national focus seems less urgent. Peripheral regions are less isolated than they were, and more ready to turn and look beyond their national borders. There is no indication of a desire to embrace supranational, quasi-federalist structures (the finely-balanced Maastricht referendum result in September 1992 is evidence enough of that), but it seems as if many in the French regions find the freedom to set their own development priorities, in concert with their neighbours and in direct communication with the institutions of the European Community, highly congenial.

That, of course, is the view from the regions. As we have seen, however, there is not much evidence that the decentralisation reforms of the 1980s have brought about fundamental change in the nature of political representation. Neither is it clear that post-Maastricht moves on the part of the European Community/European Union to give a certain prominence to the regional dimension in its decision-making processes are met with great enthusiasm in Paris. Indeed, some evidence of the French government's reaction to the newly-constituted but purely consultative Committee of the Regions can be gleaned from the decision to apportion France's entitlement of 24 seats equally between *régions*, *départements* and *communes*. The satisfaction thus afforded to the representatives of the smaller units is matched only by the disgruntlement of the *présidents de région*. Nonetheless France, with its existing forms of regional administration, remains well-placed, in marked contrast to, for example, the United Kingdom, to take advantage of the various European structural funds, the importance of which to the economies of Member States has increased substantially in recent years. In the early 1980s the main purpose of decentralisation had been seen as the modernisation of the state, the promotion of economic redistribution and local democracy. A decade and a half later, however, it is the regeneration of the market economy in partnership with private entreprise that is its main goal, with regions becoming increasingly competitive in the search for European Union funding (Loughlin & Mazey, 1995). It can be reasonably surmised, then, that European integration, with its accompanying concept of subsidiarity, will enhance the profile of the French regions, both economically and politically. It remains to be seen whether, in time, there will be a corresponding reduction in the influence of *la République indivisible*.

References:

Battye, A. & M.-A. Hintze (1992), *The French Language Today*, London, Routledge.
Benko, G. & A. Lipietz (eds) (1992), *Les Régions qui gagnent: districts et réseaux: les nouveaux paradigmes de la géographie économique*, Paris, PUF.
Braudel, F. (1986), *L'Identité de la France: espace et histoire*, Paris, Arthaud-Flammarion.
Brongniart, P. (1971), *La Région en France*, Paris, Colin.
Budd, L. (1997), 'Regional Government and Performance in France', *Regional Studies*, 31, 2 (April), 187-192.
Burke, E. (1905), *Reflections on the French Revolution*, London, Methuen.
Club Jean Moulin (1968), *Les Citoyens au pouvoir: 12 régions, 2000 communes*, Paris, Le Seuil.
Dayries, J.-J. & M. Dayries (1978), *La Régionalisation*, Paris, PUF.
Drevet, J.-F. (1988), *1992-2000: les régions françaises entre l'Europe et le déclin*, Paris, Souffles.
Fourastié, J. (1979), *Les trente glorieuses: ou la révolution invisible de 1946 à 1975*, Paris, Fayard.

Gravier, J. F. (1947), *Paris et le désert français*, Paris, Portulan.

— (1970), *La Question régionale*, Paris, Flammarion.

Gruber, A. (1986), *La Décentralisation et les institutions administratives*, Paris, Colin.

Lafont, R. (1967), *La Révolution régionaliste*, Paris, Gallimard.

— (1971), *Décoloniser en France*, Paris, Gallimard.

Lanversin, J. de, A. Lanza & F. Zitouni (1989), *La Région et l'Aménagement du territoire dans la décentralisation*, 4th edition, Paris, Economica.

Lévy, M. L. (1982), *La Population de la France des années 80*, Paris, Hatier.

Loughlin, J. & S. Mazey (eds) (1995), *The End of the French Unitary State? Ten Years of Regionalization in France (1982-1992)*, London, Frank Cass.

Mény, Y. (1974), *Centralisation et décentralisation dans le débat politique français (1945-1969)*, Paris, Pichon & Durand-Auzias.

Merriman, J. M. (ed.) (1982), *French Cities in the Nineteenth Century*, London, Hutchinson.

Newhouse, J. (1997), 'Europe's Rising Regionalism', *Foreign Affairs*, 76, 1 (January / February), 67-84.

Perrineau, P. (1987), *Régions: le baptême des urnes*, Paris, Pedone.

Schmidt, V. A. (1990), *Democratizing France: the political and administrative history of decentralization*, Cambridge, Cambridge University Press.

Uhrich, R. (1987), *La France inverse: les régions en mutation*, Paris, Economica.

Weber, E. (1977), *Peasants into Frenchmen: the modernization of rural France 1870-1914*, London, Chatto & Windus.

Zeldin, T. (1977), *France 1848-1945*, Oxford, OUP.

Belgium: a new federalism

Peter Wagstaff

Introduction

> There is a comparison to Belgium within Europe: Czechoslovakia, the nation that split in two at the end of 1992. Animosity between the two Belgian factions appears far worse than that between Czechs and Slovaks. And culturally they have much less in common than the Slovak tribes, one being Germanic, the other Latin
> (*The European*, 1-4 July 1993, p. 8).

The comparison of Belgium with former Czechoslovakia appears striking, even bizarre, but an explanation is not hard to find, even if the underlying reasons which prompted it do not bear close scrutiny. It can be argued that, to a far greater extent than any neighbouring country, the Belgian state was an artificial creation, accommodating the desires and interests of the Great Powers following the revolution of 1830. Statehood and nationhood were not, therefore, co-terminous, despite the best intentions and efforts of the ruling élites to awaken a sense of national unity. Throughout its modern history there has been an ever-present threat of fracture or rift between the various elements that make up Belgium. The nature of these elements, and of the rifts which threaten them, is not as clear-cut as might be suggested by a superficial familiarity with a state apparently sandwiched between, and polarised by, the Netherlands to the North and France to the South. The relationship between two language communities, the Walloons and the Flemings (French and Dutch-speaking), is of course a major and constant feature of debate, but it represents only one of a number of significant faultlines in Belgian national life, any of which could, and from time to time do, call into question the unitary nature of the state. Indeed, there seems to be a semi-official recognition that Belgium's natural condition is one of change, 'the gradual but unavoidable transformation of a strictly centralised State entity into a regional or federal State system' (Senelle, 1987, p. 8).

Belgium has in recent decades made substantial moves in the direction of a regional federalism not unlike that which characterised the territory before the state was created. This process is unusual in that cultural differences make themselves felt in the acute form of a linguistic division which reveals not only fears of domination by one language group over another, but also a reversal of dominant roles brought about by economic circumstances. The analysis of this 'gradual but unavoidable transformation' will begin with an examination of the historical and cultural features which have set

Belgium on its present course. This will be followed by a survey of economic factors which have transformed the relationship between the two main constituencies in Belgian life and, therefore, the political landscape. Finally, I shall address the question of the succession of political and constitutional settlements which, over the last quarter of a century, have been put forward in response to the increasingly deep schisms in Belgian society.

Two Communities: Historical and Cultural Features

The geographical and physical limits of a Belgian state are not obvious. 'Natural' frontiers, such as the River Scheldt in the North, or the Ardennes-Eifel hills in the South, do not, on the whole, coincide with the political boundaries, while in the North-West the plains of Flanders merge imperceptibly with those of Northern France. It is, therefore, hardly surprising that external commentators on the origins of Belgium generally concur with the view that the country was 'an artificial creation of the great powers' (Fitzmaurice, 1984, p. 418), 'the most contrived country in western Europe' (Huggett, 1969, p. 1). From its origins in the aftermath of the Congress of Vienna, the new state was seen as 'a bulwark against France [...] a fortress on France's northern border' (Kossmann-Putto, 1987, p. 40). Sources closer to the Belgian State, however, view the matter differently, while pre-empting the sort of observations outlined above:

> Abroad the question often arises whether Belgium is not an artificial State which owes its existence to the striving for political and military equilibrium on the part of France, Great Britain, Prussia and the Danubian Monarchy.
>
> Nothing could be further from the truth [...] The North and South Low Countries (i.e. the present-day Kingdom of the Netherlands and the present-day Kingdom of Belgium) formed a political entity, created during the 15th and 16th centuries, which consisted of various principalities and had, through the genius of the Dukes of Burgundy, been made into a particularly prosperous economic and political unit
>
> (Senelle, 1987, p. 7).

These comments are representative of many attempts to construct an a posteriori normality and legitimacy for the fledgling state. The influences on Belgian origins are, however, legion. Split between French and Germanic influence (West and East of the River Scheldt) at the Treaty of Verdun in 843 AD, the territory witnessed the rise of the duchy of Brabant and of the counties of Flanders and Hainaut in feudal times. Its period of greatest cultural dominance came with the rule of the Dukes of Burgundy in the 14th and 15th centuries. The Low Countries as a whole were allied by royal marriage with the Hapsburg empire at the end of the 15th century. Hapsburg domination lasted for a further 100 years, when the seven northern provinces, under the title of the 'United Provinces', gained independent status, and the southern provinces passed under Spanish dominion. A century and a half of intermittent strife, in which France was a major actor, led to the Treaty of Aix-la-Chapelle in 1748, when the territory of Belgium passed into Austrian hands. Further conflict led to French invasion and, finally annexation. From 1795 until 1815 French control imposed a

governing élite, the impact of whose presence would be felt for generations. The allocation of the territory to the Kingdom of the Netherlands at the Congress of Vienna in 1815 was followed by a period of intense religious and political struggle leading to the revolution against the regime of William I in 1830 and the subsequent formation of the Belgian state with the drafting of a Constitution in 1831. No unitary state had existed on this territory prior to this date. What preceded it was a loose, quasi-federalist grouping of provinces enjoying a considerable measure of autonomy under successive regimes (Logie, 1980).

Thus between 1795 and 1831 the foundations for the modern Belgian state were laid. At the same time, however, the main elements of tension and conflict were being incorporated into the structure. The arrival of the French armies in 1792, heralding annexation three years later, led to the imposition of a jacobine centralisation which was to survive the 15 year period of Dutch rule from 1815. Liberal and Catholic opinion combined to produce sufficient anti-Orangist and anti-Protestant feeling to oust the Dutch in the revolution of 1830, and the Constitution of 1831 enshrined the principle of a constitutional monarchy at the head of a unitary state with a parliamentary system of government. Even in the first flush of statehood, however, the fissiparous tendencies of the new Belgium were noted by a contemporary historian: 'In Belgium, there are parties and provinces, but no nation. Like a tent set up for one night, the new monarchy, after sheltering us from the storm, will disappear without trace' (Nothomb (1834), in Hasquin, 1982, p. 22). The inaccuracy of that prediction does not negate the underlying implication that some form of conflict or opposition was the most likely condition of the new state.

The imposition of a unitary, centralised structure was matched by a policy of unilingualism in all but name which took no account of the existence of two juxtaposed language groups. While the majority of the population was Dutch-speaking, located in the Flanders provinces in the north of the country, the newly-created state was, to all intents and purposes, francophone, with its legacy of bourgeois French-speaking élites and a governmental and administrative structure on the French model. The Constitution offered, in theory, freedom in language use, but laws and decrees were all promulgated in French and, although 'Flemish' translations existed, only the French texts had legal validity. This unilingual reality was not, however, a main source of discord in the early years. It was only as the 19th century progressed that traditional cleavages on religious and social grounds were replaced by division along linguistic lines (Wils, 1993).

Control of the newly-constituted Belgium lay with a political class dominated by the traditional opposition of Catholics and Liberals, who alternated in power for much of the 19th century. The predominance of the French-speaking sector of the population was compounded, for at least half a century after independence, by an electoral system based on property qualification – by and large it was the francophone bourgeois élites who participated in the electoral process. In 1846 Belgium contained 2.5 million Dutch speakers, 1.8 million French speakers, and a mere 45,000 qualified electors. The exclusion of almost the entirety of the Flemish population was compounded by the effects of demographic changes evident as early as the late 19th century and persisting up to the present day. The Walloon population, as a proportion of the whole, reached a peak in the 1880s, when the success of the Industrial Revolution in the southern coalfield, combined with the progressive impoverishment of the economy of Flanders, led to southward migration. From then on, however, the balance started to alter, in terms of both demographic trends and cultural and linguistic assertiveness. Partial reform of the constitution in 1892-3 initiated a process of extension of voting rights to all adult males by 1920-1 (universal male and female suffrage for national elections was not obtained until 1948).

In parallel with these political developments, which tended towards an enfranchisement of the numerically dominant but politically impotent Flemings, changes in the status of the two language groups were gradually introduced. Small, piecemeal reforms, such as the imposition of the Dutch language in court proceedings in the Flemish part of the country from 1873, and a similar imposition on government authorities in that region in 1878, followed five years later by the introduction of some teaching in Dutch at secondary level, also in the Flemish sector, culminated in the major reform of 1898 guaranteeing equal status for Dutch and French at national level. The movement, then, was away from a French unilingualism applied indifferently throughout the country, and towards a limited bilingualism. From the point of view of the Walloons, this movement was perceived less as a fair adjustment of national priorities than as an assault upon francophone interests, particularly since it coincided with a sustained period of economic depression for the southern part of the country and a fall in the Walloon birth-rate. The response to the assertion of Flemish identity tended, therefore, to be a move in the direction of federalism, at least at the level of intellectual debate, if not at grassroots level. In 1912 Jules Destrée spoke of the need for 'a Belgium created from the union of two independent peoples, brought together precisely because of that mutual independence' (Supplément au *Soir*, 15 December 1992, p. 2). The first of many compromises is implicit in the law of 1921, which made central national administration bilingual, while local administration was made monolingual within the two language groups. The apparently simple demarcation between unilingual and bilingual groupings masks a number of complex issues. While it had long been clear that the Dutch-speaking community occupied the northern half of the country and French-speakers occupied the southern half (with a small community of German speakers in the East), no attempt had been made formally to delineate these two areas in precise geographical terms.

The de facto existence of a francophone community bordering France in the South, and of a Dutch-speaking community in the North, bordering Holland, received official sanction only in the mid-1960s, since when language divisions have been clearly marked on the administrative map of Belgium. The complexity of allegiance is revealed in the status of Brussels, posing an additional difficulty, and one which has proved perhaps the most intractable throughout this entire debate for, while the capital is situated in Flemish territory, the language of its governing élites is French. Further, the growth in the city's population, and the consequent encroachment of largely francophone migrants on surrounding communities has led to tension and, ultimately, to the status of a separate region with its own complex and evolving linguistic identity (Witte *et al.*, 1984).

The designation of two unilingual regions for the country as a whole, separated by an official language line, dates from 1932, although formal legal status was only afforded in 1963, and consolidated by further legislation in 1970 in an attempt to deal with the anomalous situation of Brussels. As a result of this legislation, there is now formal definition of the language regions. The Dutch-speaking region comprises the Provinces of West Flanders, East Flanders, Antwerp, and Limburg, together with the districts of Louvain and Hal-Vilvorde which are part of the Province of Brabant (i.e. the northern part of the country). The French-speaking region comprises the Provinces of Hainaut, Namur, Luxembourg, and Liège (except the eastern part) and the district of Nivelles which is part of the Province of Brabant (i.e. the southern part of the country). The German-speaking region comprises the eastern part of the Province of Liège. The bilingual region of Greater Brussels comprises the 19 *communes* of the urban area of which the capital is a part (Senelle, 1987, p. 10). The relationship between these various regions, and their current status in terms of political and administrative responsibility, will be examined in greater detail in the third section below. The measures outlined reflect an acknowledgement of the cultural and linguistic reality concealed behind the image of the unitary state, and of the inevitable abandonment of the ambition for a nationwide bilingualism that would successfully mask the divisions.

There have been few occasions in the history of Belgium when the cohesiveness of the state has seemed attainable. Most obviously, perhaps, the wars of 1914-18 and 1939-45 might be seen as a rallying-point for national unity in the face of aggression or the violation of neutrality. And yet it was the resentment of the (largely Flemish) foot soldiers against their francophone officer corps in the trenches of the First World War that gave a boost to the Flemish movement. A generation later the early release of Flemish, but not Walloon, prisoners of war by the Germans, and the ambiguous, possibly collaborationist stance of the King, Leopold III, led to equally strong resentment in the French-speaking community, culminating in the strike of 1950 in protest against the King's return, and his forced abdication in favour of his son, Baudouin (Aron, 1977).

The political forces which have dominated Belgian life since the 19th century fall broadly into three categories – Social Christians (of Catholic origin), Liberals, and Socialists. All three were unitary in origin, each having a single structure for the entire country. The modifications to the Constitution in 1892-3 and 1918-19 which brought about the establishment of universal suffrage, accommodated the continuation of political division on religious and social lines until well after the Second World War. The Catholic Party became the *Parti Social Chrétien* (PSC) in 1945 and occupied one side of the political divide. The anti-clerical vote was split on a class basis between the old-established Liberal Party (founded in 1846), which became the *Parti de la Liberté et du Progrès* (PLP) in 1961, and the Socialist *Parti Socialiste Belge* (PSB), dating from 1885 (Senelle, 1965, p. 43). For the first half of the 20th century and beyond, the domination of the political scene by these three traditional parties was almost total. Between the end of the First World War and the early 1960s, when the language conflict took a sharper turn, the Social Christians, the Liberals, and the PSB took, on average, more than 88 per cent of the parliamentary vote, and were in fact the only parties to participate in government, with the brief exception of the Communist Party for eight months immediately after the Second World War (Mughan, 1983, p. 437). The growth of parties representing the interests of one or other linguistic community, mostly in the years following the Second World War, led the three main parties to split along similar lines, reflecting the Flemish-Walloon division. As a result, cross-party regional solidarity is probably as strongly delineated as the national political cleavages. The laws of 1962-3 which traced the demarcation line between linguistic regions were intended to defuse the political element of the linguistic dispute. The gains in that direction were short-lived, however, since animosity between communities escalated continually until, in March 1968, the government was brought down by the controversy surrounding the proposed transfer of the French-speaking section of the University of Louvain to Walloon territory. The Social Christian Party subsequently relinquished its unitarian stance, splitting into two to become the PSC (of French expression) and the *Christelijke Volkspartij* (of Dutch expression). This scission was prompted, in part at least, by the arrival on the scene of a number of *communautaire* parties (i.e. those dedicated to the interests of specific linguistic groups). Throughout the post-war period, the tide of nationalism, particularly Flemish, led to the formation of such parties. The *Volksunie* movement was founded in 1948, followed shortly thereafter, in predictable counter-reaction, by the *Front Démocratique des Francophones* (FDF) and by the *Rassemblement Wallon* (RW).

The balance of power within Belgium, then, has shifted substantially in recent generations. The Flemish population, always numerically superior and increasingly dominant in economic terms, has gradually come into its own, undermining the hegemony of the influential French-speaking minority. This prompted at first an attempt to legislate for an officially bilingual country and subsequently the recognition of the reality of two distinct linguistic communities evolving along separate if parallel lines. The result has been that the concept of the unitary state, and of a notional bilingualism within it, has been under severe pressure. Throughout the entire period of Belgian statehood, the political and cultural forces outlined above have been sustained and given direction by a range of complex and often conflicting economic pressures.

Two Communities: Economic Factors

The 19th-century Industrial Revolution, based on coal, began earlier in Belgium than elsewhere in continental Europe. Exploitation of the massive Borinage coalfield which ran from the French border through Mons, Charleroi and Namur to Liège and beyond, in Walloon territory along the Sambre-Meuse valley, gave Belgium a faster rate of industrial growth during the mid-century than either Great Britain or Germany. While the coal and iron ore deposits in the southern, francophone, part of the country at first ensured a thriving economic base, the Dutch-speaking North was for a long time unable to match this achievement. The Walloon region retained its dominant position in the Belgian national economy well into the 20th century although, as a proportion of the country's industrial workforce, the Walloon population had started to decline by the 1880s. This can be partly explained by a natural process of industrial and technological development. The primary activities of coal and ore extraction, together with basic iron and steel production, were slowly complemented by the next stage of industrial development, with the production of non-ferrous metals and chemicals in the coastal zone around Antwerp in the Flemish North. Exploitation of the mineral resources of Belgium's Congo colony also tended to concentrate activity in the privileged ports and their hinterland. This meant growth and the beginnings of prosperity for Flanders, which boasts the Belgian coastline in its entirety (a mere 65 kilometres). By the early years of the 20th century, Flemish industrial production was rising (admittedly from a very low base) more rapidly than Walloon production, which itself reached a peak around 1930. From that point onwards, the Brussels-Antwerp axis was the focus for dynamic growth in industrial sectors such as vehicles and petrochemicals, flourishing into the 1950s and beyond. In later years, the difficulties experienced by traditional heavy industry, such as cheap imports and falling demand for coal and steel, affected the Walloon Sambre-Meuse coalfield even more seriously than the neighbouring Nord-Pas de Calais of northern France, and by 1982 the Walloon share of the Belgian industrial workforce had fallen to 27.7 per cent (Thomas, 1990, p. 38). By contrast, the gradual build up of manufacturing activity in Flanders brought with it a growth in regional self-confidence and, in its train, an assertion of the rights of the Dutch-speaking population.

It should be noted that it is industry and commerce which have been the motors of the Belgian economy. The role of agriculture, for so long a dominant activity in France, is tiny – less than 5 per cent of the active population is engaged in farming, and its contribution to the economy has fallen steadily, from 6.5 per cent of GDP in 1960 to 1.8 per cent in 1990 (*L'Etat de l'Europe*, 1992, p. 510).

Demographic trends only served to reinforce the shift of influence from North to South – the population of Wallonia has remained stagnant while the birth-rate of the Flemish areas has risen substantially throughout the 20th century. In the period from 1920 to 1982, the Walloon population grew by a mere 350,000, while the Flemish population increased by 1.91 million (Thomas, 1990, p. 41). As population growth is translated into the distribution of parliamentary seats, the growth in Flemish influence is a natural concomitant of a rising birth-rate. Most importantly, it is in the active population that the shifts are most clearly revealed. Between 1947 and 1961, there was an increase in the proportion of the active population in all the northern provinces except East Flanders, whereas all the provinces of Wallonia, without exception, suffered a decline (Huggett, 1969, p. 83). It is hardly surprising, therefore, that political divisions are also highlighted and modified by these demographic changes. The Catholic-based PSC traditionally draws substantial support from the Flemish population; its influence has therefore been consolidated by Flemish ascendancy. If Flanders is predominantly Catholic, then Wallonia has a more secular culture, providing support for the PSB and the small Belgian Communist Party, as well as the PLP. As well as marking one more division between the communities, this religious/anti-clerical opposition is a reflection of the industrial climate as it has developed over the years. Belgium as a whole has one of the highest population densities in Europe, with over 250 inhabitants per square kilometre and, while it is Antwerp, Ghent and Brussels which are the most crowded areas now, during the early period of Walloon industrial ascendancy in the second half of the 19th century the population was highly concentrated in the south. This concentration provided vast reserves of cheap labour, a situation which accounts for Marx's description of Belgium as 'a capitalist's paradise' (Huggett, 1969, p. 28), but also explains the appeal of the anti-clerical parties of the left, notably the PSB. Labour militancy in the Walloon coalfield, leading to strikes and production stoppages, has not been reflected in the Flemish north, where the Catholic church has normally been quick to condemn militant action and where industrial peace has therefore tended to reign. These factors explain, in part at least, the lack of investment in the stagnant Walloon industrial economy, and the relative success of Flanders in attracting incoming investment in industry. The post-war period, and particularly the 1960s, saw, therefore, a rapid growth in the economy of Flanders, in terms of manufacturing industry and the service sector. The change in the relative importance of the industrial and service sectors in the period 1960-1990 is instructive – a gradual decline in industry's contribution from 40.9 per cent of GDP in 1960 to 30.1 per cent in 1990 has been mirrored by a rise in service sector activity from 52.6 per cent to 68.1 per cent.

Brussels became a centre for service sector employment, enjoying a period of enormous expansion in this area from the late 1950s onwards. The obvious stimulus was the creation of the European Economic Community and the selection of Brussels as the seat of the Commission and the Council of Ministers. The magnetic effect on foreign companies and countries alike was striking, with the establishment of more than 200 embassies and the offices of over 1,100 international organisations (Thomas, 1990, p. 39). The very success of Brussels as a focus for service sector activity has, however, contributed to serious problems elsewhere, particularly in Wallonia, which lacks the range of provincial urban development to compete with the capital. In comparison with Flemish territory, with its long-established urban tradition rooted in the Middle Ages (Bruges, Leuwen), Wallonia's urban development is based essentially on the industrial expansion of the 19th century, and its towns consequently tend to lack the facilities which would permit diversification and growth.

In the modern period, therefore, three factors combine to produce a need for radical change to the institutional framework of the Belgian state. First, the growth in the confidence and economic power of the Flemish area, to match its demographic ascendancy. Second, the concomitant weakness of the Walloon constituency, and its failure to manage successfully its reconversion from a heavy industrial past. Third, the status of Brussels and its surrounding area – economically buoyant through an explosion in tertiary activity, yet occupying an ambiguous position both culturally and linguistically. Brussels of course is not the only part of Belgium to present an anomalous and potentially conflictual situation. At various points along the dividing line that separates the Flemish and French-speaking regions, arrangements have had to be made to protect the interests of language minorities from either side, with similar arrangements in place for the benefit of minority German-speakers in those areas adjacent to the Germanophone zones in the east. Of particular note is the situation of the enclave known as the Fourons, on the northern fringe of the French-speaking province of Liège. Responsibility for the district was transferred in 1963 to the Flemish province of Limburg, provoking dissatisfaction and resentment which provided fertile ground for French and Flemish nationalists, leading to a bitter dispute and the downfall of the centre-right government in 1987 (Poole, 1987).

It was these factors, and the growing acerbity of relations between the various parties concerned, that led to the succession of revisions to the Belgian constitution from the 1960s onwards.

Constitutional Revision: Towards a Federal Structure

The signs of Flemish-Walloon dissension were sufficiently evident in the years following the Second World War to prompt the setting up of a research centre, the Centre Harmel, in the search for 'a national solution to social, political and juridical problems in the Walloon and Flemish regions' (Supplément au *Soir*, 15 December 1992). This ensured that informed and influential debate on the subject, and agreement on reform measures, were carried on in an extra-parliamentary forum in the years that followed, with the Parliament exercising an a posteriori legislative role. The deliberations of the Centre Harmel took account, throughout the 1950s, of both the growth in Flemish economic power and influence and, in response, a gradually increasing determination on the part of the Walloon community to protect its declining interests by pressing for a more federalist state structure. The Walloon population found such a structure all the more attractive as it watched the Flemish parliamentary majority direct state investment towards the Flemish economy. By the end of the 1950s, the Centre Harmel had proposed that the two language communities should be granted cultural autonomy, and the language laws of 1962-63 enshrined this principle. In addition, these laws brought a new element into play by sanctioning an equivalence between cultural and territorial divisions. The linguistic dividing line between Flemish and French-speaking territories henceforth appears on the map as a physical feature.

In the three decades that have followed this recognition of a match between linguistic and geographical divisions, further constitutional revisions have been made in response to a complicated interplay of political demands and compromises. The 1970 revision saw the introduction of two new concepts which were intended to meet the differing demands of the two major language groups. In response to Flemish requests, the principle of the *Communauté* emerged. This represents an entity formed according to cultural (i.e. linguistic) criteria. At the same time, and in response to Walloon desires for an element of socio-economic autonomy, the principle of the territorially-based *Région* was brought into being. At once, each of the three *Communautés* (French, Flemish, and German-speaking) assumed responsibility for cultural and linguistic affairs. The development of a framework for the operation of the *Régions* was, however, a much more complex matter, since it proved impossible in the early stages to resolve the question of Brussels, with its superimposed language communities. A further revision of the Constitution in 1980 merely extended the responsibilities of the *Communautés* in areas such as health care, while at the same time giving a theoretically autonomous status to the French and Flemish *Régions* (Mabille, 1986). By the revision of 1988, however, a much wider range of responsibilities was devolved to *Communautés* (in particular education) and *Régions* (public works and transport). By this stage, something like 30 per cent of the national budget was allocated at the level of *Communauté* and *Région*, although this substantial budgetary responsibility was not yet matched by an equivalent democratic responsibility – regional and *communauté* councils were not directly elected. Perhaps the most significant advance lay in the fact that the Brussels region was defined, at least provisionally, within the limits of its 19 *communes*.

A notable feature of the constitutional revisions which have taken place over the last quarter of a century has been the increasing momentum of change. The process has been a gradual but ineluctable movement in the direction of a federal state, and the most recent revision of the constitution, the *Accord de la St Michel*, debated at the end of 1992 and passed into law on 14 July 1993, has been hailed as the necessary, decisive step towards federalism. There is an unfortunate irony in the fact that this watershed legislation was closely followed by the death of Baudouin, *roi des Belges* (King of the Belgians) and institutional symbol of the attempts to forge and to maintain a united kingdom. The implications of the 1993 revision are far-reaching, the new Article 1 of the Constitution states that 'Belgium is a federal state', although in some respects perhaps not quite as far-reaching as the proponents of a fully-fledged federalism would wish. The status of the three *Régions* (Wallonia, Flanders, and Brussels-Capital) has been confirmed and consolidated, with a range of responsibilities specified. Regional development, transport and public works, housing, economic and employment policy, environmental matters, and relations with the *communes* and *provinces*, all depend on decision-making at regional level. By the same token, the responsibilities of the three *Communautés* (French, Flemish, and German-speaking), have been specified, and relate largely to the cultural sphere. Cultural heritage, radio and television, tourism, policy for youth, education, inter-*communauté* cooperation, health and family policy, form the bulk of *Communauté* responsibilities, together with the question of language use in administration and education. This represents an extension, but no fundamental change, to the immediately preceding situation. The major problem for the *Communautés* has been, and remains, the division of responsibilities within the Brussels-Capital region itself which, with its two-language mix, fits uneasily into the *Communauté* framework. ⌐Resolution of the problem depends on the success of a complicated set of institutions designed to ensure even-handedness while acknowledging the specific identities of the two language communities, which must co-exist on the same territory.⌐ Regional matters are dealt with, as for the two other *Régions*, by a directly elected regional council, while community matters are subject to a complex tri-partite regime composed of commissions for French, Flemish, and joint responsibilities. However, the most radical element in the 1993 revision concerns the alterations to Belgium's bi-cameral parliamentary system. The effect of these changes is to bring to an end the previous equality of power between the Senate and the *Chambre des Représentants*. Henceforth, the *Chambre* has wide-ranging decision-making powers, in the realm of the budget, ministerial responsibilities, and motions of confidence, which it can exercise without regard to the Senate. True, certain areas continue to require joint *Chambre*-Senate agreement, such as constitutional revisions, international treaties, and the delegation of powers to the European Union and other international bodies. The role of the Senate, on the other hand, becomes more generally consultative, with the power to propose legislation, to which the *Chambre* can react as it thinks fit. The composition of the Senate is, at least in part, indicative of its role as a federal chamber. The previous number of 184 senators is reduced to 71, 40 of whom (25 Dutch, 15 French-speaking) are directly elected. A further 21 (10 Dutch, 10 French, 1 German-speaking) are nominated by the three *conseils de communauté*, with another 10 (6 Dutch and 4 French-speaking) co-opted by the previous category. Thus, 31 senators derive their status from the *Communautés*, reflecting the federal intentions of the newly-constituted senate. Overall, however, the ratio of Dutch- to French-speaking senators (41 to 29, or 58.6 per cent to 41.4 per cent) is very closely linked to the ratio of the two groups within the population as a whole. Paradoxically, this does not reflect the normal mode of representation in an institution structured on federal lines. The expectation is of a system based either on equal representation for the constituent parts (e.g. the U.S Senate) or one in which the smaller constituencies receive compensatory representation (e.g. the German *Bundesrat*). There is one further feature of the newly revised constitutional arrangements which tends to belie the federalist intentions. Unlike most federal states, Belgium retains a judicial system which, in terms of its organisation and its day-to-day working, is common to the state as a whole, rather than being allowed to evolve separately within the *Communautés*.

Conclusion

It would seem, therefore, that Belgium has perhaps not advanced quite as far down the road towards the federal state, even with the most recent constitutional revision, as commentators have claimed. Nonetheless, many features of federalism are in place, and it is clear that their introduction is seen as a bulwark against the quasi-separatist tendencies of the two major language communities. In addition, difficult economic circumstances since the end of the 1980s have seen a concomitant rise in the threat of political extremism, although an increase in the percentage of the vote gained by the extreme right Vlaams Blok, mainly in Antwerp, and of the Walloon counterpart the Front National, between the 1991 and 1995 elections (to 10.1 per cent) still left them with only 13 seats in the 150-seat federal chamber. In this context, too, federalism is felt to offer an alternative to sharply focused antagonisms. At the same time as the institutional structure of the Belgian state has been subject to these changes, successive governments, whether of the centre-left under M. Martens between 1988 and 1991, or of the centre-right under M. Dehaene since 1991, have sought to promote the status of Brussels as European capital, rather than simply as the home of the European Commission and bridgehead for countless international organisations and companies. Provisions in the Maastricht Treaty for regional and community executives to be represented at the Council of Ministers are seen as a means of defusing regional antagonisms by allowing them to be expressed in a wider, supra-national forum. A Belgium which aspires to a form of federalism within a Europe where similar aspirations are being voiced, however tentatively, is one which acknowledges the existence of the faultline between Europe's Latin and Germanic cultures within its own borders, and yet hopes for a reconciliation of its own conflicts on a wider stage. This does not, however, prevent leading Flemish politicians voicing from time to time an interest in the creation of a 'Greater Flanders' which would embrace northern Belgium and the southern Netherlands, having dismissed the Belgian state as 'an accident of history' (*The Economist*, 20 September 1997). The motivation for such talk is the argument, reminiscent of that heard in Northern Italy in relation to the South, that a prosperous and enterprising Flanders is obliged to subsidise a crumbling Walloon economy (Murphy, 1995, p. 92). The response from some Walloon politicians at least has been talk of 'rattachisme' – merging southern Belgium with France.

It will be clear from the above that, given the uniquely dominant influence of the language question on Belgian national life, there are few conclusions to be drawn that are directly applicable to the regional debate within the European Union as a whole. It is hard to argue with the view that 'there will never be any danger of Belgium's being cited as the classic example of a federal state' (Witte, 1992, p. 115). Indeed, even within Belgium itself, the nine pre-existing regional units have been largely eclipsed by the major three-way division (four Dutch-speaking, four French-speaking, and Brussels-Capital with its own complex formulation). The assertion that, faced with an apparent choice between separatism and a unitary state, Belgium may find 'a middle way based on federalism within a Europe of the regions' (Fitzmaurice, 1996, pp. 267-268) is seductive. It takes no account, however, of a systemic fragility which leaves the state vulnerable to acute intercommunal suspicions and recriminations. Most recently these have taken the form of accusations of widespread cronyism and corruption in the political class and the judiciary, and reactions to events such as the multiple child killings and paedophile scandal in the summer of 1996, events whose reverberations would, in a more firmly based state, be restricted to and contained by the criminal justice system but which, in this instance, have provoked speculation on the integrity of the state itself. As Downs suggests, in the wake of the 1995 elections, 'federalism may have been achieved, but cleaning up the corruption and relieving the underlying social ills that breed support for separatist and anti-democratic political forces will not allow Belgium's federalists much time to rest on their laurels' (Downs, 1996, p. 174).

However, the impact over the long term of the strengthening of strictly regional representation at various levels of European Union deliberations, whether at the Council of Ministers or the Committee of the Regions, may well encourage the emergence of a more measured, and less politically charged, atmosphere than has been evident in recent months and years.

References:
Aron, R. (1977), *Léopold III ou le choix impossible: février 1934 - juillet 1940*, Paris, Plon.

Downs, W. M. (1996), 'Federalism achieved: The Belgian Elections of May 1995', *West European Politics*, 19 (1), 168-175.

Dumont, G.-H. (1991), *La Belgique*, Paris, PUF.

The Economist, 20 September 1997.

Feron, F. & A. Thoraval (eds) (1992), *L'Etat de l'Europe*, Paris, La Découverte.

Fitzmaurice, J. (1984), 'Belgium: Reluctant Federalism', *Parliamentary Affairs*, 37, 418-433.
— (1996), *The Politics of Belgium: A Unique Federalism*, London, Hurst.

Frognier, A. P., M. Quevit & M. Stenbock (1982), 'Regional Imbalances and Centre-Periphery Relations in Belgium', in S. Rokkan & D. W. Urwin (eds) , *The Politics of Territorial Identity: Studies in European Regionalism*, London, Sage.

Hasquin, H. (1982), *Historiographie et politique: Essai sur l'histoire de la Belgique et la Wallonie*, 2nd edition, Charleroi, Institut Jules Destrée.

Huggett, F. E. (1969), *Modern Belgium*, London, Pall Mall Press.

Kossmann-Putto, J. A. & E. H. Kossmann (1987), *The Low Countries: History of the Northern and Southern Netherlands*, Flanders, Flemish Netherlands Foundation.

Logie, J. (1980), *1830: De la régionalisation à l'indépendance*, Paris, Duculot.

Mabille, X. (1986), *Histoire politique de la Belgique: facteurs et acteurs de changement*, Brussels, Centre de recherche et d'information socio-politiques.

Mughan, A. (1983), 'Accommodation or Diffusion in the Management of Linguistic Conflict in Belgium', *Political Studies*, XXXI, 434-451.

Murphy, A. (1995), 'Belgium's Regional Divergence: Along the Road to Federation', in Smith, G., *Federalism: The Multiethnic Challenge*, London, Longman, pp. 73-100.

Poole, A. (1987), 'The Fourons: a microcosm of Belgium's linguistic problems', *The Linguist*, 26 (2), 52-56.

Senelle, R. (1987), *The Reform of the Belgian State*, Vol. IV, Brussels, Ministry of Foreign Affairs and External Trade (Memo from Belgium No. 196).

Thomas, P. (1990), 'Belgium's North-South Divide and the Walloon Regional Problem', *Geography*, 326, 75 (1) (January 1990), 36-50.

Wils, L. (1993), 'Belgium on the Path to Equal Language Rights up to 1939', in *Ethnic Groups and Language Rights (Comparative Studies on Governments and Non-Dominant Ethnic Groups in Europe, 1850-1940*, Vol. III), Dartmouth, New York University Press.

Witte, E. *et al* (1984), *Le Bilinguisme en Belgique: le cas de Bruxelles*, Brussels, Editions de l'Université de Bruxelles.

Witte, E. (1992), 'Belgian Federalism: Towards Complexity and Asymmetry', *West European Politics*, 15 (4), 95-117.

Regionalism in the Netherlands

Bernard O'Sullivan and Denis Linehan

Introduction

The Netherlands, like other smaller members of the European Union, has been an ardent supporter of the European integration process. The furthering of integration within the Union has historically been seen as a means to moderate the power of its larger northern neighbour, while offering ample economic opportunities to reinforce the Netherlands's position as the 'Gateway to Europe'. The effects of integration have also had important consequences for regional governance within the Netherlands. In common with European Union Member States such as Belgium, Italy and Spain, the Netherlands has recognised the need to develop policies to provide opportunities for regions to direct economic development within their own territories. Such policies have been based first on an understanding that the regions in the Netherlands now exist as interdependent entities within the national and European economy, second on a conviction that regions should take an active role in managing their own economies and, third, on an understanding that central government can no longer be wholly responsible for rectifying regional economic imbalance within the Netherlands. Reflecting on the increased significance of European integration, this chapter identifies the central drivers of change that have encouraged central government to support the development of economic regionalism in the Netherlands. Key phases in the development of the administrative structure of the Netherlands and the evolution of Dutch regional policy are examined. Within this context it is argued that recent Dutch policies highlight important aspects of successful regional strategies in Europe which have sought to achieve optimum benefit from European integration. The chapter illustrates these trends further by examining how the Dutch Province of Limburg, exposed in the past to profound restructuring problems with the decline of its traditional industry, re-invented its economy within the spaces provided by change in Dutch regional policy and the process of European integration.

Dutch Governance and the Sub-National Tiers

Dutch government has often been described in a somewhat contradictory fashion as a 'decentralised unitary state'. Known as the system of Home Administration, this unique fusion of centralised and decentralised powers consists of three layers of administration – Municipal, Provincial and National. The structure is guaranteed under the constitution of 1848, whose principal architect, Johan Thorbecke, envisioned the modern state as a complex whole, composed of dynamic but interdependent parts.

A student of organic state theory and a sharp critic of the French tradition of centralised government, fundamental to his thinking was the value of the 'organic' or 'natural' interdependence of Municipal, Provincial and National layers of government. Within his design, autonomy was granted to the Municipalities and the Provinces, but constitutional checks were provided whereby the Provinces enjoyed supervisory powers over the Municipality, and in turn, National government could execute control over the Province. This system, it was judged, provided the means of conflict resolution at the appropriate levels.

Clearly, Dutch administration today is far more complex, given the emergence of more governmental actors, such as specialised administrative bodies and corporatist institutions, but Thorbecke's rationale of co-governance is indicative of the Netherlands's consensual approach to societal matters as a whole and retains a powerful influence in many aspects of Dutch civil society. Critics are divided however as to the benefits of this system of co-government (Hendriks, 1997). It has been argued that the existing administrative structure no longer relates to the geographical and economic realities of the Netherlands today and makes the running of the state over-bureaucratic. In particular, the provincial tier of government has been the subject of much debate and criticism. A key point of contention is associated with the perceived administrative gap between the Province and Municipalities in a period of expanding urban growth, notably in the Randstad. To remedy this problem, the amalgamation of Provinces and the creation of city regional authorities has been proposed. The idea of provincial amalgamation emerges from the belief that the Provinces in the Netherlands are too small and that larger administrative units would be the equivalent of German *Länder* and offer greater economies of scale. In addition, it has been argued that the creation of city-regional authorities would allow urban cores greater flexibility in meeting the demands of competitive growth and change within an integrating Europe. These plans are particularly attractive to the more powerful municipalities such as Rotterdam which have always been keen to reduce what is often regarded as Provincial interference in their development policies.

As yet such developments have not come to fruition and, in a move which stimulated significant controversy, the Provinces have retained their power and indeed in association with increased European Integration, have further developed their position as the legitimate administrative tool to interpret national and European policy and adapt it to the demands of their territories.

The Evolution of Dutch Regional Policy

The recent emergence of the regional and provincial role in managing economic restructuring must be located first within the context of post-war regional planning and secondly within economic and institutional shifts in the Netherlands which have paralleled European integration in the last two decades. Historically, regional policy in the Netherlands was conducted in a redistributive manner, using a largely Keynesian model of state intervention. Central government allocated resources selectively to reduce regional inequalities and balance economic growth across the country. Regional assistance was first directed to peripheral rural areas in the North and was extended in

THE NETHERLANDS

Groningen
•
GRONINGEN
Leeuwarden
•
FRIESLAND

Assen
•
DRENTHE

NORTH
HOLLAND

Lelystad
•
FLEVOLAND

Amsterdam
•

OVERIJSSEL

SOUTH
HOLLAND

Enschede
•

• Leiden
• The Hague

Utrecht •

GELDERLAND

UTRECHT

Arnhem
•

Rotterdam
•Dordrecht

Nijmegen
•

ZEELAND

GERMANY

NORTH BRABANT

Middleburg •

Eindhoven
•

LIMBURG

BELGIUM

Heerlen
•

Maastricht
•

0　　　　　50

kilometres

the mid-1960s to areas undergoing de-industrialisation, such as the southern provinces of Limburg and Brabant. In the 1970s, the negative effects of over- development in areas such as the Randstad promoted further policies to encourage the dispersion of industries from urban centres to less populated areas. The Selective Investment Regulation Policy, for example, placed levies on new firm formation in the Randstad and strict procedures were introduced for firms wishing to locate in Rotterdam.

As in other European countries, the impact of the first oil crisis in the mid-1970s brought such policies into question and concern was raised that these restrictive policies served to undermine the competitiveness of the national economy as a whole. In addition, it was argued that while regional policy in this form initially improved the distribution of economic activity, it failed in the long term to promote local economic growth, since many of the branch plants directed from the Randstad to peripheral areas closed as the European economy contracted.

The political demands generated by the economic restructuring that followed from this crisis had significant consequences for the Netherlands, in terms of the kind of institutional responses and new policies stimulated within the state. Traditional instruments of regional policy, developed during the 1960s and 1970s, proved unfit to deal with the problems of structural change. In the course of the 1980s they were progressively abandoned, heralding the wider retreat of government intervention in the economy and a disillusionment with the 'blueprint' planning which was a hallmark of Dutch regional planning in the post-war period. Centrally-controlled restrictions on industrial location were relaxed and central subsidies cut back or re-orientated. Calls for alternative policies soon emerged and these centred on the needs for greater local autonomy as a means to meet the demands of industrial change. In this way, the demands and eventual changes in regional policy that occurred in the Netherlands corresponded to aims of economic policy elsewhere in Europe, resulting in the reorganisation of institutional instruments at the regional level. For some commentators the coherence of this response, notably the new reliance upon regional actors to manage the consequences of change, has been explained as the result of comparable institutional responses effected in capitalist economies under what is widely understood as Post-Fordist economic conditions (Harvey, 1989). Thus in the Netherlands, Dijkink has argued that 'the retreat of government is not only a temporary phenomenon, compelled by fiscal entrenchment and economic recession, but it is also considered an essential characteristic of the post-industrial condition' (Dijkink, 1995).

The new approach to regional policy was enshrined in *The Fourth Policy Document on Physical Planning in the Netherlands* introduced in the late 1980s. As in other European states, its rationale was to move from a static form of 'blueprint' planning to a more flexible 'process' planning. This policy change evolved in a period of economic uncertainty, increased European integration and the ongoing re-structuring of the Dutch economy. The role of central government shifted towards encouraging regional authorities to meet the demands of change and an emphasis was placed on supporting the economy by concentrating on key national economic strengths, rather than seeking an even distribution of development. The central feature of the policy was the emphasis placed upon encouraging collaboration between Provinces and Municipalities and the private sector. Private investment in regional projects was promoted through the use of innovative public-private partnerships, reducing state investment, but retaining a role when large and costly projects were involved. The report identified nine nodal points in reorganising the competitive profile of the country. These included the four urban areas of the Randstad (Amsterdam, Rotterdam, The Hague and Utrecht) as well as Groningen, Enschede, Hengelo and Maastricht. Their selection was based on the quality of education, labour, telecommunications and geographical position in relation to national and European networks. The role of central government was to assist in removing transport bottle-necks, reinforce the international competitiveness of the cities and strengthen their positions as regional centres. Clearly, the Randstad and the rest of the western Netherlands was an important focus in terms of protecting Dutch international competitiveness, but all regions were encouraged to 'choose their own lane for the next century'(Clement, 1991).

The result was a decentralisation of regional policy, involving critically not just the re-allocation of resources but also of economic decision-making powers. Through this shift, central government has actively promoted sub-national government to respond independently to the national and international economic challenges, including increased economic integration, faced by the Netherlands. The new policy's primary objective was to maximise the economic efficiency of all its regions and thereby increase a region's individual contribution to national economic growth. 'Offensive policies concentrating on strengths are pursued, rather than defensive policies focused on territorially concentrated weaknesses'(Bachtler & Clement, 1992). The provincial authorities were encouraged to produce alternative contingencies for economic development and to develop rapid and effective solutions for existent or emergent problems by utilising new funds made available by central government. An investment budget of 2 million Guilders was, for example, allocated to each province to assist in the establishment of Small Manufacturing Enterprises. In addition, a special fund was established to support firms seeking new and specialised production and market niches. An Investment Grant Scheme allocated grants to provinces for a four year term on merit and this was extended for another two years on the basis of effective utilisation of earlier funds. The Municipalities also took on new responsibilities in the fields of new firm formation, the transfer of specialist knowledge from the public to private sectors and in city marketing and promotion campaigns.

An important feature of the new policy was that extra discretionary funding was provided for those Provinces which produced innovative and structured policies in response to economic change. These discretionary incentives in turn stimulated Provincial governments to become increasingly inventive in mobilising concrete support for the regional economy through the incorporation of key social partners in the business and trade union arena, all of whom were regarded as an integral part of the regional development policy formulation. In this way central government enhanced the invention of new measures and policy concepts through the institution of a subsidy system for local states measures, so that the latter had to compete for flexible funding. 'Thus it was not necessarily the economically weakest regions but those putting forward the most sophisticated development projects, which were given support' (Tommel, 1992, p.117).

A further feature of the new regional policy was the reinvigoration of the Regional Development Corporations created first in the early 1970s to assist in the execution of central government policy. These bodies were now encouraged to adopt a more pro-active role in promoting a 'bottom-up' approach to regional development and currently provide support for companies in the form of strategic advice on technological, financial and export issues. Operating as private firms with majority government shareholdings, they finance themselves through loans on capital markets with powers to take shares in existing firms and assist start-up firms. For example, the Gelderland Development Corporation offers its clients various services in cooperation with the central Ministry of Economic Affairs, the Provincial Government of Gelderland and the main municipalities in the Province. Employing 12 specialised project managers, the corporation provides information and advice to local and overseas companies on industrial and office location, available incentive schemes, technological innovations and export possibilities. It also offers financial services and can purchase shares or provide loans as venture capital for new firms. It should be noted that this policy was directed towards ensuring regions took advantage of opportunities presented by increasing European integration but was not created because of the availability of co-financing European stimuli. The Netherlands has traditionally received little in way of European structural funding; however, such funding has played a subsidiary role in areas such as Limburg.

Within these contexts, the nature of regionalism within the Netherlands can be said to have changed. The combination of shifts in National government policy and the re-launch of the European project, effectively energised the regions to create their own agendas within the national and European arenas. The restructuring of state and economy, particularly in the context of European integration, has in turn restructured the nature of regionalism. Traditionally conceptualised in terms of separation from the state, it can be argued that these new networked institutional arrangements at the regional level constitute part of a new development in political regionalism in Europe in which the Netherlands has readily participated. Dutch regional policy abandoned the Keynesian model of intervention and utilised flexible policies where central government has adopted an indirect role as a partner and facilitator of development. Responsibility to organise and mobilise the resources at the regional tier has been placed on networks of interests including Provincial and Municipal government, the private sector and the trade unions.

Critically, the Dutch heritage of 'co-government' has been a valuable resource in a period of increased European integration where the sub-national tiers of government have greater opportunities for action. Unlike the United Kingdom for example, which has only recently come to terms with need for dynamic regional institutions, the Netherlands has been fortunate in already having such institutions in place. While politically the transitions have not been without their controversies, particularly with the larger, more powerful Dutch municipalities, this 'decentralised unitary state' has managed to capitalise on recent trends in the European economy with some considerable success. This chapter continues by examining in more detail how the province of Limburg has adapted to these new responsibilities and how it has reoriented its economic profile to manage the opportunities offered by increased European integration.

Limburg – 'The Dutch Wedge Into Europe'

The province of Limburg provides an illustrative example of the dynamics of decentralised regional policy in the Netherlands. Situated in the south-east of the Netherlands, surrounded on three sides by Belgium and Germany, Limburg is positioned between the growth poles of the Dutch Randstad, the Ruhrgebeit in Germany (Dortmund and Essen) and Brussels and Antwerp in Belgium. The Province has a unique cultural history, which has helped shape the identity of the region within the Netherlands. In the latter part of the 19th century the national borders in this area were relatively insignificant. There was continuous mobility of workers, students and investors in the area and a familiarity with three languages among the inhabitants. In the immediate post-Second World War period however, the application of centrally coordinated development strategies in the individual countries served to pull the region apart.

As with a number of other traditional industrial regions within Europe in the 1970s, such as South Wales, Limburg's overt dependence on coal mining left it exposed to the collapse of that sector. By the early 1980s the industry was effectively ruined, leaving widespread unemployment, an uncompetitive economic infrastructure and environmental degradation. This economic upheaval was alleviated by a concerted restructuring policy guided by the Provincial government within the framework offered by *The Fourth Policy Document on Physical Planning in the Netherlands* . In keeping with many of the key tenets of economic regionalism recently witnessed in Europe, such as in Nord-Pas-de-Calais and the North-West of England, the execution of Limburg's new regional strategy provides an example of the new approach to Dutch regional policy based on public-private partnership and one influenced heavily by economic opportunities offered by European integration.

The wider objectives of the regional strategy were to redefine Limburg as a service-oriented Euro-Region. Provincial planning sought to capitalise on the region's locational advantages relative to the European market through new investments in transport, distribution infrastructure, telecommunications and tourism. The institutional responsibility for this new approach to regional development is clear cut. The Province provides the necessary infrastructure and incentives. The Regional Development Corporation (NV Industriebank, LIOF), provides information and investment advice to companies, while the four Chambers of Commerce offer legal and planning assistance to small companies. In addition there is a series of government sponsored agencies involved in a number of labour-market and training initiatives, such as Limburg Centre for Company Training (LCB) and the Centre for Education and Labour (COA). All these bodies work together under a framework shaped by the provincial government.

The social and cultural infrastructure of the province was enhanced through the founding of a new university and international research centres, and a European policy typified by the European summits held at Maastricht. Both LIOF and Maastricht City Council embarked on a major marketing campaign designed to re-invent Maastricht as an international city – as a convention centre, a centre of communication (in the heartland of the Rhine-Meuse Euroregion), a major tourist attraction and an international centre for education research and development. This strategy, in addition to support from the European Union, has led to the establishment of some 25 major international institutes in the city, including The European Institute of Public Administration, The European Centre for Work and Society, The European Centre for Development Policy Management, Teikyo Europe (University for the Japanese Community in Western Europe) and The European Institute for International Communication. The internationalisation of Limburg was also designed to attract inward investment. Again this developmental strategy was coordinated by LIOF in partnership with the provincial government and the municipalities. LIOF acted as a point of information and mediated between governmental institutions on behalf of interested companies. Utilising the selling points of developed assets such as the Technoport Europe business park located near Maastricht airport and the Venlo tradeport (an advanced rail and road transportation base), Limburg has attracted inward investment from over 250 foreign companies, including Mitsubishi, Matsui, Mobil and Memorex Telex.

As part of its Europeanisation strategy, the creation of the Rhine-Meuse Euro-Region provides an example of the provincial government availing itself of opportunities provided by increased European integration. The Euro-Region straddles the national frontiers of three countries, including the cities of Maastricht, Heerlen, Aachen, Liege and Hasselt. The launch of the single European market created further opportunities for the dismantling of economic barriers and helped promote cooperation between firms and institutions in the region. The Limburg provincial government has been at the forefront of developing the Rhine-Meuse project, with support from central government which highlighted the opportunities that Limburg could exploit vis-à-vis Aachen and Liège. An important factor was the success in winning funding under the INTERREG programme in the early 1990s, a European Union initiative which aimed to facilitate cross-border collaboration. The fund provided investment of up to 14 million Dutch Guilders over a four year period. The projects that received European Union funding included the establishment of regional information databases and the development of cross-border joint ventures. These projects include a cross-border industrial park developed by the city councils of Aachen and Heerlen and the MHAL project that identifies common physical planning priorities in the region, such as the improvement of communications linkages between the cities, the extension of the TGV network to Genk/Hasse and the development of advanced distribution facilities within the region. European funding of this nature provided a unique opportunity to support the transition from a declining region to a prosperous player in European markets.

Clearly, the support from central government and the European Union has played a significant role in Limburg's economic transition, but the province's new-found success has been built predominantly on the abilities of the actors within the Province to marshal their resources and project a new identity for the Province. A critical factor in this identity formation was the influence of the then Provincial Governor, Dr J. Kremers. His leadership provided the vision of the new identity for Limburg, whereby the province would become a European region, throw off its peripheral status and become 'the Dutch wedge into Europe'. Other marketing slogans such as 'Southern Europe begins at Limburg' and 'Limburg: the Balcony of Europe' were used by the Provincial Governor to exalt this new identity for the region, as well as an ambitious plan for new public architecture. However his call for a new identity was underwritten by a rigorous political strategy. Through strong contacts within the ruling Christian Democrat governments during the 1980s, Kremers secured significant amounts of central funding. The decentralisation to Limburg of central government departments such as civil service pensions and the central statistics bureau was a result of his lobbying. In addition, while Kremers provided the new identity and the money to turn his image of the new Limburg into reality, his strong leadership style ensured there was no dissension internally within the Province. He ensured that the cities, employer organisations, trade unions and chambers of commerce followed a unified provincial policy. He ensured the Province was the only tier of government negotiating with The Hague with regard to finance and development policy, arguing that the Province needed one voice. Some have criticised this political strategy as autocratic, but as A.H. Verheijen, of the LIOF, remarked, 'the period of reconstruction offered the Provincial government the chance to get everyone's noses pointing in the same direction. It is certainly true that Brussels and The Hague are also prepared to support this line, however their support has to be earned. Without solidarity in the region, meaning a strategic plan with clear-cut agreements, no support would have been forthcoming'.[1] The realisation of such strategies indicates the importance of the subtle negotiations required between culture, politics and economy in developing successful regional formations.

Conclusion

In a wider European context the change in Dutch regional policy mirrors changes in many other European Union Member States, as regional tiers of governance are offered the tools and 'strategic space' to develop pro-active economic policies within an integrating European economy. It has been argued that the Netherlands, due to its inheritance of 'co-government' has been better placed than other member countries to mobilise its regional economies accordingly. Nevertheless, the debate continues over the need to overhaul the administrative system and the provincial tier in particular. The larger Dutch municipalities fear a 'Jutlandisation' of the Netherlands if regional reform is not carried through and larger regional entities created in order to streamline decision-making and create regions of a necessary geographical scale that would be more appropriate to a 'Europe of the Regions'. While scale does have its merits, as does the need to quicken the pace of decision-making, such institutional changes are only one side of the argument. In the Netherlands the clamour for radical institutional change often masks a lack of strategic creativity, a dilemma encapsulated by a former Director General of the National Physical Planning Agency in the phrase 'we have all the ingredients, but we just don't any longer know which cake to bake'(Witsen, 1991). The municipalities would perhaps do well to observe the lessons of Limburg, a peripheral economy within the Netherlands, which addressed change through an imaginative re-interpretation of its economic profile, a regional development strategy based on public-private partnerships and a redeployment of its resources to take advantages of change in the national and international spheres. Limburg provides a salutary lesson not only for other regions in the Netherlands but also for economic regionalism throughout Europe. While national and supra-national change has provided greater opportunities for regions it also means that regions are obliged to become more creative in developing their economic agendas and use of resources. Recourse to radical institutional change is a poor substitute for strategic ability. It is in effect a step back to an older form of regionalism based on the mistaken belief that institutional change for the sake of change, coupled with greater autonomy, automatically produce superior administrative and economic performance.

Note

1.   Interview with A. H. Verheijen, Executive Manager (Beleidsmedewerker) LIOF, April 1995.

References

Bachtler J. & C. Clement (1992), 'Community Regional Policies for the 1990's', *Regional Studies*, 26 (3), 414.

Clement, C. (1991), 'Regional Economic Policy for the Next Few Years', *Tijdschrift voor Econ. en Soc. Geografie* 82 (3), 228.

Dijkink, G. (1995), 'Metropolitan government as a political pet? Realism and tradition in administrative reform in the Netherlands', *Political Geography*, 14 (4), 335.

Harvey, D. (1989), 'From managerialism to entrepreneurialism: the transformation of urban governance in late capitalism', *Geografiska Annaler*,71b, 3-17.

Hendriks, F. (1997), 'Regional reform in the Netherlands: reorganising the viscous state', in Keating, M. & J. Loughlin (eds), *The Political Economy of Regionalism*, Frank Cass, London, pp. 370-387.

Meijer, H. (1989), *The Fourth Policy Document on Physical Planning in the Netherlands*, IDC, Utrecht, Den Haag.

Tommel, I. (1992), 'Decentralisation of regional development policies in the Netherlands – A new type of state intervention?' *West European Politics*, 15 (2), 107-125.

Toonen, T. (1993), 'Dutch Provinces and the struggle for the meso' in Sharpe, L.(ed.) *The Rise of Meso Government in Europe*, London.

Witsen, J. (1991), 'Five decades, five directors: the national physical planning agency 1941-1991 – A Personal View', *Built Environment*, 17 (1), 65.

Federalism in Germany

Theo Stammen

Introduction

When the Federal Republic of Germany (*die Bundesrepublik*) was founded in 1949, its political and institutional structure included a large federalist component. Three factors contributed to this. The first stemmed from the reality of the situation at the time – most of the federal states (*Länder*) in the *Bundesrepublik* had been either freshly created or set up along existing lines by the Allied Occupying Powers in the period between 1946 and 1949 and were consequently older than the *Bundesrepublik* itself. The second factor can be traced to the prescriptive status of the 'Frankfurt Documents', which the three western military governors imposed on the German Prime Ministers (*Ministerpräsident*) of the *Länder* in the summer of 1948 as the framework for the creation of the West German Constitution, or Basic Law (*Grundgesetz*), and which stipulated that the new German constitutional order should be democratic and federalist (Currie, 1994). Finally, a third, traditional, factor came into play – in contrast to the majority of centrally-organised European nation-states, such as France or Great Britain, Germany had, throughout its recent history, always had a federal structure, reflecting its great regional variety. In the 19th and 20th centuries, different forms of federalist system succeeded one another – either confederate (*Deutscher Bund*) or federal (*Deutsches Reich*) bringing together principalities and dukedoms. The Weimar Republic, too, was a federal state which established a precarious balance of power between central power and regions, *Reich* and *Länder*, with the balance tilted in favour of the *Reich*. It was notable, also, that the National Socialist dictatorship very quickly got rid of the *Länder*, and with them the principle of a federal structure, so as to consolidate the centralist power structure of the NS system (Gunlicks, 1995, p. 220).

Federalism, then, along with the principles of the rule of law, democracy, republicanism and the social state, took its place as a fundamental feature of the *Bundesrepublik's* new constitutional structure. While this principle has consistently underpinned the political system, it has nonetheless undergone considerable change in the period since 1949.

It is worth noting, too, that the other German state to be founded in 1949, the German Democratic Republic (DDR), was a federal state, according to its first constitution, which was very much in the image of the Weimar constitution, with five *Länder* and a parliamentary chamber (*Länderkammer*). By the beginning of the 1950s, however, this original federal component of the DDR constitution was abolished in favour of a strict centralisation of state power linked to the construction of socialism.

Fourteen administrative districts ((*Verwaltungs*)-*Bezirke*) took the place of the *Länder*, and the *Länderkammer* ceased to exist as a DDR institution. Interestingly, in 1990, during the process of German unification, the re-formation of the five *Länder* in the DDR (as it still was at the time) echoed this tradition. In October 1990, these five *Länder* formally joined the Federal Republic on the basis of the unification agreement and Article 23 of the *Grundgesetz*. This increased the number of *Länder* from 11 to 16.

It is the structure of the *Bundesrepublik* – at present the only traditionally federal Member State of the European Union, apart from Austria – which is the subject of this chapter. Above all, the following three problem areas need to be addressed and analysed – the formation and development of the federal structure of the *Bundesrepublik*; the constitutional significance of federalism; and the current political importance of German federalism in relation to reunification and European integration.

The Federal State System in Germany

The basic principles of the *Grundgesetz* are set out in Article 20 – 'The Federal Republic of Germany is a democratic and social federal state.' This, in conjunction with Article 28, which states that 'the constitutional order in the *Länder* must conform with the principles of republican, democratic, and social government under the rule of law, within the meaning of this Basic Law', contains the prescriptive basis for the *Bundesrepublik's* constitutional framework. The basic meaning of these five structural principles is essentially underlined in Article 79, paragraph 3 of the *Grundgesetz*, which expressly excludes any alteration to these constitutional rules ('Amendments of this Basic Law affecting the division of the Federation into *Länder* in legislation, or the basic principles laid down in Articles 1 and 20 shall be inadmissible'). This means that not even a 100 per cent majority in the national parliament (*Bundestag*) and the representative chamber for the *Länder* (*Bundesrat*) would be able to remove the federal structure of the *Bundesrepublik* by the legitimate process of a constitutional amendment; it is not surprising, then, that there has been talk of an 'eternal guarantee' of federalism (Kilper & Lhotta, 1996, p. 79). Although talk of 'eternity' is always problematical in the political sphere, this formula makes it expressly clear that federalism is especially protected and guaranteed as a structuring principle of the *Bundesrepublik's* political and institutional order – it belongs to the permanent core of German constitutional law, the existence of which cannot be infringed in the process of European integration. However, the fact that the federal principle is guaranteed by the Basic Law is no obstacle to reform of federalism in Germany. According to the Constitution, it is entirely possible for *Land* boundaries to be revised, and for a number of *Länder* to be amalgamated into one. Indeed, this is exactly what happened in the early 1950s, when Baden-Würtemberg was created out of three smaller *Länder*. Again, during the constitutional reform that followed reunification, there were attempts to amalgamate Berlin and Brandenburg. This proposal was, however, rejected when it was put to the people in a referendum.

The principle of federalism (Article 20 of the *Grundgesetz*) determines the vertical structure of the political system. If, as seems appropriate, we include local

government, then the overall political structure of the *Bundesrepublik* can be visualised in terms of a large building with three separate but linked storeys consisting first, on the ground floor, of local government, with its municipalities (*Gemeinden*), towns (*Städte*) and districts (*Kreise*) then, on the first floor, of the 16 *Länder* each with its own constitution and political institutions, and finally, on the upper floor, of the federal state with its central state organs. The *Grundgesetz* as the constitution of the *Bundesrepublik* determines the complex architectonics of the entire building. This piece of architecture could be extended to include a fourth floor if the European Community were included, as may perhaps be appropriate in the future with the realisation of European union. The political activity that takes place simultaneously on all floors is correspondingly diverse – each level has its own political tasks and responsibilities, its own processes for the development of informed political opinion and decision-making, with appropriate political institutions which are democratically constituted and legitimised. Accordingly, there are three (four) different sorts of election – municipal, *Land*, federal (and European) elections, which are held every four or five years, according to the various electoral rules. The existence of these three political levels in the German political system means that citizens, as well as parties and organizations, have a variety of opportunites to participate in the democratic process.

The political parties are active at every level of the State, and are the most important organisations for the development of political opinion, in that they present the voters with a variety of programmes and candidates, and compete for political office. Nowadays it is only at the level of local government that it is still possible to win a political mandate without the support of a political party. At the level of the *Länder* and the federal state the parties effectively have a complete monopoly of political representation and participation.

The various strata of state organisation as a whole can be seen as a system of vertical power distribution and control. This is complemented at each level by a horizontal distribution of powers. Above all, the federal and *Land* levels operate, in association with each other, as a system of double (horizontal and vertical) distribution and control of power; therein lies one of the most important constitutional functions of federalism in Germany.

The independence and specific responsibility of local self-government as the lowest level of the entire federal structure of the State is laid down in Article 28, paragraph 2, of the *Grundgesetz*: 'Communities must be guaranteed the right to regulate all the affairs of the local community on their own responsibility, within the limits set by statute'. This right also applies to associations of communities – they too have the right to self-government within their own legal area of responsibility according to the laws. The principle of subsidiarity underpins the basic philosophy of federalism in many ways – smaller units should only cede responsibility to larger ones when a particular task or responsibility is beyond their means, or when it is necessary in order to ensure that standards of uniformity are maintained.

As fundamental as this constitutional guarantee of self-government for municipalities, towns and districts is for the State's federal structure, it has to be admitted that developments since the beginning of the *Bundesrepublik* have tended more and more to increase the dependency of municipalities and towns on the *Länder* and the federal state, particularly in financial terms. This financial dependency has meant that *Land* and Federation (*Bund*) respectively, through the exercise of their constitutional rights, have been able severely to restrict the political capabilities and freedom of action of the towns and municipalities. The financial burden on these local authorities, which are responsible for welfare payments, has risen considerably during the current crisis of the welfare state prompted by growing unemployment; the result has been fewer resources for their other activities, for which money can no longer be found. As yet, no way has been found to solve this problem. Nonetheless, municipal self-government remains the basis of Germany's federal structure, following its reorganisation after far-reaching reforms of the municipalities in the early 1970s. Yet it is rarely at the forefront of political, media or academic concerns, which invariably focus on *Bund* and *Länder* and especially the distribution of power and jurisdiction between them. It is clear that, in spite of the fact that it is at the local level that the citizen can have the greatest say in his or her immediate political environment, municipal elections arouse comparatively less interest than those held at *Landestag* and *Bundestag* level. It is apparent that, even at municipal level, voters believe *Land* and *Bund* politics have a greater impact on them personally than municipal politics.

The central problem of federalism has two aspects – first the distribution of responsibilities and functions between the Federation (*Bund*) and individual *Länder*; and second, the problem of finding a fair and appropriate means of allowing the *Länder* to participate in federal policy decisions through a special federal organ.

When the *Bundesrepublik* was founded there was much debate about the right way to approach these two constitutional problems. In both cases several possible solutions were considered. The question of the distribution of responsibilities and functions revolved above all around the question of finance. Two variations had already been tried, with mixed success, at earlier stages in German history. First, at the time of Bismarck, when the *Reich* as the central power had remained financially dependent on the *Länder*. Second, during the Weimar Republic, when the *Länder* were, in the end, disastrously dependent on the *Reich* in financial terms – disastrous in the sense that this one-sided financial relationship left the political independence and freedom of action of the *Länder* seriously weakened. The result was that, in the latter stages of the Weimar Republic (1930-33), they were no longer able to offer any significant resistance to the rise of National Socialism. This led to the simultaneous collapse of the Weimar Republic and Hitler's seizure of power. The National Socialist State that resulted was organised in a centralist mode; it therefore stood in contradiction to every tradition of German state formation and organisation.

As a result of this historical experience, the founders of the *Bundesrepublik* constitution attempted a far-reaching revision of the federal structure, in order to avoid the problems and dangers of the earlier arrangements. Above all, the way responsibilities were divided up in the fields of legislation and administration needed to be reorganised. An essential precondition for this was an improved financial framework, which would avoid the (one-sided) dependency both of the *Bund* on the

Länder as well as the reverse dependency of the *Länder* on the *Bund*, and would guarantee for both actors as large as possible a measure of financial autonomy.

As far as the division of legislative responsibility is concerned, German federalism does not follow the concept of strict division (as in the USA) but rather a combination of functions. There is a tri-partite distinction between legislation which is exclusive to the *Bund* and *Länder* respectively, framework legislation for the *Bund*, and concurrent legislation affecting both the *Bund* and *Länder*. What this distinction means in detail is itemised and defined in Chapter VII of the constitution. So for example the subjects of the *Bund's* exclusive legislation in Article 73 are itemised in just as much detail as the subjects of concurrent legislation. The list of subject matter is long and the sphere of concurrent legislation has been enlarged constantly over the years, to the benefit of the *Bund*. For example, 'The Federation has the right to legislate in these matters to the extent that a need for regulation by federal legislation exists' (Article 72 of the *Grundgesetz*). In this context, the *Länder* may only legislate 'so long as and to the extent that the Federation does not exercise its right to legislate'. However, this has been the case less and less frequently in the history of the *Bundesrepublik*. The *Bund* has exercised that right just as thoroughly for concurrent legislation as in the field of framework legislation. The main reason for this development is to be found in Article 72, (paragraphs 2, 3), 'the maintenance of uniformity of living conditions'. The *Länder* have also called upon this principle in support of their activity in those areas where they can claim exclusive legislative competence, for example in the sphere of cultural and education policy (schools, universities, science, culture, etc). By setting up a standing conference of education ministers, for example, the *Länder* have created a special organ of self-coordination (without participation from the *Bund*) capable of monitoring the uniformity of standards in the field of education and culture. It was entirely logical that the *Länder* should create specialist committees (*Bildungsrat* and *Wissenschaftsrat*), for the coordination and integration of education and science policies, and that in the end the *Bund* should bring this increasingly important area within the sphere of its own framework legislation (e.g. framework legislation for universities) and within its own financial ambit. This example shows that development was constantly in the direction of a unitary state (*Unitarisierung*); hence the talk of a 'unitary federal Republic' or of 'cooperative federalism', to describe this growing tendency towards standardisation in the German federal system. The exclusive responsibilities of the *Länder* have been reduced in number, and have become more restricted in scope. Now they are more or less limited to the regulation of the police and education; and even in these areas the *Bund* has become increasingly active through its framework legislation. This transfer of power and responsibilities from the *Länder* to the *Bund* has led to the creation of new ministries, such as the Federal Ministry for Research.

However, it is notable that, as the legislative independence of the *Länder* has been reduced, there has been a rise in the responsibilities of the *Bundesrat* as the representation of the *Länder* at federal level, for the growing share of concurrent and framework legislation taken over from the *Bund* is composed, naturally enough, of subjects for legislation which require agreement because they concern the interests of the *Länder* and cannot be enacted without the agreement of the *Bundesrat*. The number of such subjects has grown from one legislative period to the next, and amounts to about 75 per cent or 80 per cent of laws – a clear indication of the considerably strengthened position of the *Bundesrat* in the legislative process of the *Bundesrepublik*.

The integration of responsibilities between *Bund* and *Länder* can also be seen in administration. Chapter VIII of the *Grundgesetz* rules specifically on 'the execution of federal statutes and the federal administration'. It is in this area that the *Länder* chiefly exercise their specific responsibilities, for the *Bund* has separate administrative responsibility for only very few spheres of activity, such as foreign affairs (diplomacy), federal financial administration, major traffic and transport administration (air, roads, water), Federal Border Guards and a few other minor areas, and similarly also in the federal armed forces and their administration. In general, Article 83 deals with the ways in which administration operates, 'the *Länder* shall execute federal statutes as matters of their own concern insofar as this Basic Law does not otherwise provide or permit'. Essentially, the task of the *Länder* is to run an administration to undertake the functions of the executive. Finally it should be remembered that, through close cooperation of *Bund* and *Länder* ('cooperative federalism') a growing number of 'joint tasks' have appeared, which the *Bund* contributes to at the level of planning and finance. This means that the *Bund* plays a part in the implementation of *Länder* tasks if these are relevant to the country as a whole and if the *Bund's* contribution is necessary for the improvement of general living standards. The most important spheres of activity for these joint tasks are construction and extension of universities, and structural improvements to the economy and agriculture of the region.

The *Bundesrat*, as the organ of federal legislation, has been able significantly to increase its political influence through the institution of these joint tasks, while as a result the *Länder* have, for their part, suffered a loss of political independence. This has become a significant political factor, especially in a situation where, as at present, the majority in the *Bundesrat* is at odds with the *Bundestag*. In this case, the opposition which is in a minority in Parliament can use its majority in the *Bundesrat* to continue its opposition, and to force the government to compromise, for it is simply not possible for laws to be passed without the agreement of the *Bundesrat*. When the two chambers are unable to come to agreement on a bill, then the mediation committee, composed equally of members of the *Bundestag* and the *Bundesrat*, is called in, and has, so far, managed to find a compromise in most disputes between government and opposition. However, with the approach of the 1998 General Election, the successful work of the mediation committee, with its significant projects for governmental reform, has come to a standstill. The committee has also been unable to reach agreement on tax, pensions and welfare reforms, and this has meant that political debate has virtually ground to a halt. As a result, the *Bundesrat* has become the effective seat of opposition politics, which from a constitutional point of view was certainly not intended.

The system of division of responsibilities between *Bund* and *Länder* requires an appropriate division of tax revenues. In this respect, the financial constitution (or *Finanzverfassung*) represents a central element of the federal state constitution as a whole (cf the *Grundgesetz*, Chapter X, 'Finance' ('*Das Finanzwesen*') (Art. 104a - 115)). The *Finanzwesen* contains various rulings of fundamental significance for the federal structure: first in relation to legislative responsibility in financial and taxation matters. Since the reform of 1969, the *Bund* has very much the major role here; nonetheless, federal tax laws as a rule need the agreement of the *Länder* in the *Bundesrat*.

The allocation of tax revenue (vertical redistribution) is of central importance. Article 106 sets out in detail which taxes and state income the *Bund* and *Länder* are entitled to. It is interesting that the most important taxes (income, corporation, and sales (VAT) tax) are apportioned jointly to the *Bund* and the *Länder*: income and corporation tax are divided equally between *Bund* and *Länder*, while the distribution of sales or value added tax is determined by federal law (with the agreement of the *Bundesrat*), and the ratio may be varied according to circumstances. This has happened recently in connection with the so-called 'solidarity agreement' ('*Solidarpakt*') for the new *Länder* and has altered in favour of the *Länder*. During the reunification process attempts were made to adjust the constitutional relationship between the *Bund* and the *Länder* to the advantage of the *Länder*. No progress in this area has been possible, however, particularly where the new *Länder* are concerned, because of the current difficult economic and budgetary position.

We need, though, to remember that the districts (*Gemeinden*), which are the lowest level of the federal system and which would otherwise receive only a very small provision from their own taxes (mainly trade taxes), receive a share of the income tax revenue of the *Länder*; the precise size is determined by federal law for the entire *Bundesrepublik*. Since there are often considerable differences between the *Länder* with regard to size, economic strength and consequently tax yield, there is a process of 'horizontal financial redistribution'. In this way, it is intended to ensure 'a reasonable equalisation between financially strong and financially weak *Länder*' (Article 107, paragraph 2 of the *Grundgesetz*). Accordingly, a distinction is made between those *Länder* which are financially weak and therefore entitled to redistribution in their favour, and those which are financially strong and are therefore obliged to contribute; equalisation between the two is carried out on a legal basis. It is significant that, in the course of the history of the *Bundesrepublik*, variations in economic development since 1949 have caused substantial shifts in the relationship between the two categories. Some *Länder* which once belonged to the contributing group have now joined those in receipt of redistributed income; other *Länder* which initially benefited from redistribution now find themselves obliged to contribute. Regional economic policy has had differing effects here; old, classic industrial regions (such as the *Ruhrgebiet*) have declined in importance, while new industrial regions (such as Baden Würtemberg, around Stuttgart) with the most modern electronic technology, have emerged. The original economic discrepancy between North and South has long since evolved into a distinction between South-West and North-East; today it is Baden Würtemberg, Hessen and Bavaria which are at the summit of economic power in the *Bundesrepublik*. Since unification, the five new *Länder* are grouped together at the foot of the table. Their economic and financial plight has made it impossible for them to be admitted immediately to the federal system of financial redistribution. The essential support which they require was provided until 1995 by the 'German Unity Fund' ('*Fonds Deutsche Einheit*'), set up by the *Bund* and the *Länder*. Unsurprisingly, the federal financial constitution remains very much a subject of controversy (Exler, 1993). With the political situation in a constant state of flux, the material and financial challenges, big and small, to the joint policies of *Bund* and *Länder* are subject to endless change. At the moment, with the process of German unification underway and faced with substantial economic and financial problems, they seem particularly large and, for a long time to come, will continue to represent probably the greatest challenge to the *Bundesrepublik*'s federal system.

The second controversial point concerns the institutional framework within which the *Länder* can participate in federal policy-making, and here too there is a number of possible variations. On one side, the American Senate principle could be followed, according to which the population of the *Länder* would democratically elect a certain number of representatives (senators), who would constitute a senate as a second parliamentary chamber. This would then participate in the federal state decision-making and legislative process alongside the representative body. Alternatively, it would be possible to follow the *Rat* principle, under which the governments of the *Länder* form a sort of second chamber through which their interests are looked after at federal level.

The founding fathers of the *Bundesrepublik* constitution came down in favour of the *Rat* solution not least on the grounds of old German traditions and experiences. The *Bundesrat* as the federal organ of the *Länder* is not elected, but consists of representatives of the *Länder* governments, whose task it is to look after the interests of the *Länder*. Each *Land* therefore has a number of votes according to its size ('Each *Land* has at least three votes: *Länder* with more than five million inhabitants have four, *Länder* with more than six million inhabitants five. *Länder* with more than seven million inhabitants have six' (Article 51, paragraph 2 of the *Grundgesetz*)), and when there is a ballot these votes may only be cast en bloc. The *Reich* under Bismarck and the Weimar Republic had operated under the same *Rat* principle. At the assembly which decided on the form of the constitution, (the *'Parlamentarische Rat'*), only a small minority of members was in favour of the American Senate principle, which therefore had no chance of being adopted.

It is extremely hard to assess whether, overall, this was a sensible decision; the criteria to be applied vary according to circumstances – with the criterion of efficiency in mind, the *Rat* principle has to be favoured, to the extent that it is to be expected that the *Länder* governments, with the support of their ministerial bureaucracy, can work more effectively and efficiently in the interests of their *Länder* than a few representatives, whose loyalty as a rule is primarily to their party and who therefore pursue national rather than regional goals. Under the criterion of democratisation, the senate principle finds equally strong favour, since senate members are directly elected by the inhabitants of their *Land* and therefore can claim direct democratic legitimacy for their task.

If we compare the achievements of the representative body of the German *Länder* (the *Bundesrat*), constituted along federal lines, with other federal second chambers in Europe or elsewhere, then its efficiency and quality cannot be overstated. Furthermore, it must be stressed that the *Bundesrat*, in the course of developments from 1949 to the present day, has been able substantially to enhance its political position and significance. Above all, its share of the *Bund's* legislative activity has constantly grown. There are fewer and fewer parliamentary bills which are not declared by the *Bundesrat* to be in need of its agreement. This tendency has at times been so much in evidence that the federal constitutional court (*Bundesverfassungsgericht*) has had to limit the power of the *Bundesrat* to impose the requirement for agreement on bills, in favour of the federal government.

We are faced therefore today with a paradoxical situation with regard to both basic decisions; as far as the regulations concerning the responsibilities of *Bund* and *Länder* are concerned, this development has clearly worked to the benefit overall of federal responsibilites. The *Länder*, and particularly the *Landtage*, the *Länder* parliaments, have declined in importance as political decision-making authorities. As for the involvement of the *Länder* at the federal level, through the *Bundesrat* with its *Länder* representatives, the result has been a considerable expansion of the need for agreement and therefore a noticeable strengthening of the role of the *Bundesrat*.

Recently, in connection with German unification and European integration, the *Länder* have adopted common initiatives with the aim of extending still further the significance of the *Länder* in federal politics; this aim is served particularly by the recently adopted revision of article 23 of the constitution, which is intended to increase the participation of the *Länder* in federal decisions concerning Europe.

The Constitutional Function of Federalism today

We have seen that the re-introduction of federalism as a basic principle of the *Grundgesetz* was brought about by considerations which are partly traditional and historical, partly constitutional, and partly modern.

Federalism has always been at the heart of Germany's political system; this is reflected in the varied patterns of state formation on German territory since the Middle Ages. Furthermore, the attempts at nation-state integration in the 19th century were only possible within the existing federal structure.

Attention was focused particularly strongly on constitutional considerations after the Second World War because of the unfortunate fate of the Weimar Republic and the experiences with the totalitarian National Socialist state. The weak position of the *Länder* in relation to the *Reich* in the Weimar system, the way in which they were rapidly brought into line during the NS seizure of power, as well as the subsequent definitive elimination of the federal structure of the German *Reich* and the setting up of a totalitarian dictatorship – all these were experiences which stimulated constitutional discussions in the post-war period and also promoted the reinstatement of the federal component in the formation of the German state. Against this background of contemporary history, the re-establishment of the *Länder* and of federalism should be seen as part of a many-sided reflection on constitutional matters, the most important aspects of which are those which relate to democracy and the rule of law.

The democratisation of Germany, as conceived by the victors in the Potsdam agreement of August 1945, gave prominence to the view that German involvement with democracy and democratic rules should be allowed to develop from the ground up, that is first at the lower, communal level, then at the level of the *Länder* and only then at the level of the state as a whole. This meant that this concept of democracy formed the basis for a federal state organisation at three levels – communes, *Länder*, and *Bund*. Democratisation was thought of essentially as an educative process, which would be developed from elementary and easily understood situations in the communal sphere, evolving gradually towards more complex and difficult situations at *Länder* and state level. The new federal structure was enormously influential in the way that it determined the various levels of this educative process.

The division of Germany which rapidly followed steered this process of democratisation in different directions and along different paths in West and East Germany. The early GDR (1949-52) was indeed a state constructed on federal lines, with a constitution closely modelled on that of the Weimar Republic; however, once the goal of the construction of socialism along Soviet lines had been proclaimed, then this original federal concept lost all meaning and was replaced by a centralist-bureaucratic concept.

In West Germany, this process developed differently; here the federal structure was of fundamental constitutional importance for the formation of the new *Bundesrepublik*, both under pressure from the Western Occupying Powers and also because of the desire for democracy on the part of West German politicians. The fact that Germany had traditionally had a federal system certainly played an important supporting role, but it was not in itself a decisive one. At the same time preoccupations with a constitutional framework loomed large – because of the experiences, mentioned above, with the totalitarian dictatorship of National Socialism. There can be no doubt that a federal system has a strengthening and intensifying effect on democracy through its various political levels; seen in this light, there is a positive correlation between the principles of federalism and democracy in the *Grundgesetz*. In practice the federal system, with its various political levels (communes, *Länder*, and *Bund*), brings about a multiplication of democratic institutions and corresponding opportunities for democratic participation. At each of these levels, there are parliaments and governments and other political authorities, for which appropriate staff have to be recruited. Democratic elections make their contribution too, and political parties gain considerably in areas of activity and influence through the state's distinctive federal structure, since they do not just take part in the process of political opinion-forming nationwide but are also active at communal and *Länder* level. It is quite frequently the case that parties which are part of the opposition at national level have charge of majorities and governments at *Länder* level, and vice versa. This produces a substantially deeper and more beneficial integration of the political parties in the political system and, at the same time, provides opportunities for politicians and political new blood to participate in the political process at all three levels. Most prominent German politicians in the post-war period including, for example, former Federal Chancellors, had as a rule been able to gain political experience at the communal or *Länder* level. It is also very much the case that federal politicians, after careers as members of parliament or even ministers have then taken on leading positions, for example as Prime Minister of a *Land* or as Mayor of a large city. This means that the various political levels of the federal system are inter-changeable one with the other, and there is an exchange of personnel in both an upward and a downward direction.

This intensification of democracy also affects the way in which ordinary citizens have an opportunity to participate in the political process. To the extent that he or she is simultaneously citizen of a town, a *Land*, and the *Bundesrepublik*, he or she has a civic responsibilty at all three levels (and now also at the European level). In concrete terms, that means that, over a period of four or five years, there will be opportunities to vote four times; that he or she will be able to have a say in the make-up of parliaments and (indirectly) of governments four times every four or five years. In the past, the citizens of the *Bundesrepublik* have readily accepted these opportunities. They have however attributed varying degrees of importance to the different elections; as a result, communal elections attract fewer voters, elections to the *Länder* parliaments attract more, with clearly the largest turnout being for elections to the federal parliament. In this latter case, the turnout is of the order of 90 per cent; it is only very recently that the number of abstentions in all forms of election has risen dramatically, as a result of disenchantment with politics and parties. So far it is unclear whether these abstentions can be put down to politically motivated protest rather than to a growing political apathy. Whichever it is (the 1998 elections will probably provide a better indication of the real reasons), there can be no doubt that the federal structure of the *Bundesrepublik* gives a significant impetus to democratic participation, in terms both of breadth and depth. Therein lies one of the most important constitutional functions of that structure. In this respect federalism has proved its worth in Germany.

There are, however, other constitutional factors to consider, not least the principle of the state as embodiment of the rule of law. This relates principally to the theme of the division and control of power, functions which, it can be argued, are much enhanced by the state's federal structure. The accuracy of this thesis is easily verifiable – in brief, the classic horizontal division and control of power is completed by an equally effective vertical division. In effect this means that the traditional horizontal division between Parliament and government and the courts, between legislative, executive and judiciary, is extended and deepened by the federal structure's multiple vertical division between communes, *Länder* and *Bund*.

Where the principle of the rule of law is concerned, this system provides not only a double division of power but also an enlargement and safeguard of individual freedoms in the face of state power. This double division increases the opportunities for protection under the law. Political despotism can be therefore more readily, and more rapidly, kept in check. This is also the main reason why dictatorial or totalitarian regimes always ensure the abolition or destruction of the federal system as a first priority (c.f. NS regime; DDR regime).

Federalism has proved its worth, too, in the field of culture and education policy where, as we have seen, the *Länder* have a particular responsibility. The diversity of educational and cultural provision in Germany is well known. In contrast to Great

Britain and France, the *Bundesrepublik* does not have a dominant cultural centre; whether Berlin, as capital, can take on this role in due course is questionable and may not be entirely desirable. Germany's diversity in this field has its roots in the old princely and bourgeois traditions, but has been underlined by modern federalism (it is an area in which the *Länder* compete with each other), and can be seen as a positive constitutional value at a time when the principle of multiplicity tends to be subordinated to the principle of unity (and centrality). Also, it must be remembered that this cultural diversity is reflected in the spheres of education and science, schools and universities. The pattern of university education in the *Bundesrepublik* is thus very pluralistic. The support and enlargement of universities and colleges are primarily a task for the *Länder*, although for years the *Bund* has contributed to the allocation of financial resources.

One final justification for federalism can be found in the constitutional discussions which took place in Germany in the aftermath of 1945, and it is one which has taken on a new significance in the current context of European integration. This is the principle of subsidiarity. It was in part the social teaching of the Catholic Church which served as a basis for the decision by the founding fathers of the German constitution in 1949 to enshrine subsidiarity as a constitutional principle. Specifically, the expression of this principle was found in the papal encyclical '*Quadragesimo anno*' of 1931. This declares that 'that which an individual can achieve on his own initiative and by his own ability, should not be taken away from him and arrogated instead to a function of society'. It is generally in contravention of justice that 'that which lower or smaller communities are able to achieve and put to good use should be claimed for the benefit of the wider community'. For 'each and every social activity is of course by its very nature and by definition subsidiary; it should support the constituent parts of society, but should never crush or absorb them.' Quite clearly, this general socio-political maxim is a very reasonable one, and provides an outstanding justification of the federal principle with its vertical differentiation of power. It can also offer a useful critical standard for judging the extent to which the centre may at any time legitimately intervene in areas for which individual federal entities have responsibility. It is equally clear that, in the political reality of today's *Bundesrepublik*, the implementation of the subsidiarity principle through the central state is a source of difficulty and that, in the balance between *Bund* and *Länder*, the weight has clearly tended to shift in favour of the *Bund*. This does not mean, however, that the question of subsidiarity has been settled and put to one side. On the contrary, it remains valid as a normative constitutional principle for federalism. The Maastricht Treaty, reflecting the progress towards European integration which it embodies, has given major emphasis to the principle of subsidiarity, a principle which is at the heart of federalism. It is clear, then, that the constitutional principles of democracy and the rule of law are very much in harmony with federalism, providing reciprocal strength and support.

In contrast, however, the principles of federalism and the welfare state do not fit harmoniously together; there is no denying the tensions between them, which are visible throughout the *Bundesrepublik* system. In essence, these tensions arise from the fact that it is hard to reconcile the basic values of multiplicity and uniformity on which federalism and the welfare state respectively are based. And this in spite of the fact that, as we have already seen, Article 20 of the *Grundgesetz* includes the phrase 'social [i.e.welfare] federal state'.

While federalism is founded on the concepts of plurality and multiplicity, the welfare state is based on equality and uniformity; it establishes as the primary policy objective the 'maintenance of legal or economic unity, especially the maintenance of uniformity of living conditions beyond the territory of any one *Land*'. (Article 72, 2, paragraph 3).

We have already pointed out how, over a period of time, the independence and responsibility of the *Länder* have been considerably reduced, with a corresponding increase in central state responsibility for action, as a result of the constitutional position of the welfare state. This tension appears unavoidable, for the social or welfare state seems to have a 'natural', inherent tendency towards centralism or centralisation. The increasing modern need for public welfare provision accentuates the trend to sacrifice the variety of regional and federal arrangements in public life in favour of the strict law of standardisation which the pursuit of equality demands. The notion of solidarity in the federal state as a whole is therefore under threat. The more prosperous *Länder*, which are obliged to hand over resources to the poorer *Länder* through the financial adjustment process, are inclined to reject the notion of solidarity between the *Länder* as a whole, especially in sensitive areas such as health and welfare policy (Jeffery, 1998, p.112). Until fairly recently these tendencies met with considerable resistance both in the political field and among the public.

'Nothing lasts longer than the temporary.' This saying which, until German unification in 1990, was often used to describe the political system of the *Bundesrepublik* and the stability of its constitutional structure, is no less applicable to the *Länder*. Created after the Second World War by the Occupying Powers within the context of the four occupation zones, they were for the most part accidental and artificial constructs. In hardly any instance did they reflect historical traditions or former regional or even state affiliations. The only exceptions are the two Hanseatic towns of Bremen and Hamburg as well as, above all, the Free State of Bavaria (the new *Land* of Saxony is also founded on old historical traditions); these evolved as legitimate, historical, political entities, and it is significant that they have therefore been able to create their own political identity.

The somewhat haphazard and provisional way in which the *Länder* were initially set up after 1945 led to the provision in the second of the three 'Frankfurt Documents' (1948) for a reform of the *Länder* boundaries, and also to the formulation in Article 29 paragraph 1 of the *Grundgesetz*, 'a new delimitation of federal territory may be made to ensure that the Länder by their size and capacity are able effectively to fulfil the functions incumbent upon them. Due regard shall be given to regional, historical, and cultural ties, economic expediency, and the requirements of local and regional planning.' [NB the restructuring of the *Länder* as stipulated by the constitution (permitting alterations to both the borders and the number of *Länder*) does not contradict the 'eternity' guarantee of Article 79, paragraph 3 of the constitution; this guarantee relates to the principle of the federal state, not the number of *Länder* or their individual responsibilities.]

So far, nothing has come of this restructuring stipulation, in spite of numerous declarations of intent, proposals, plans, blueprints, etc. (Benz, 1993). Only once in the 40 year history of the *Bundesrepublik* has there been a partial reform of the federalist system, in the South-West of the country – in the early 1950s three small *Länder*, Baden, Württemberg-Baden and Württemberg-Hohenzollern, located partly in the American and partly in the French sector, merged to form the new 'South-West state' of Baden-Württemberg (with its seat of government in Stuttgart). That is how it has remained to this day, in spite of numerous commissions, reports, plans and proposals. Those *Länder* which have been in existence since the foundation of the *Bundesrepublik* have proved themselves, for all their inadequacies, to be so stable and lasting that all restructuring plans, which in any case have often been based on questionable logic, have up to now remained unsuccessful. This resistance to all such plans, which invariably allowed for a (more or less drastic) reduction in the number of *Länder*, can be explained in part at least by the instinct for self-preservation displayed by many *Länder* politicians, who are naturally aware that the break-up or merging of individual *Länder* would mean the loss of political office and would therefore put their own political careers in jeopardy. As a result of this political inertia, many restructuring plans have disappeared from public discussion, and the constitutional provision set out in Article 29, paragraph 2 has remained unfulfilled from 1949 to German unification in 1990.

Meanwhile, German unification has brought about a new situation – the question of restructuring the federal regions is once more on the agenda. The re-federalisation of East Germany has provided, in addition, a good opportunity to revive the idea of restructuring the federal regions as a whole, in accordance with the requirements of Article 29, paragraph 2 of the constitution.

Nevertheless, it is fairly easy to predict that this new attempt at restructuring the federal regions, which will be contested in part with the old arguments and essentially old plans, will not lead to any concrete new reorganisation of the *Bundesrepublik*. Nothing more than a partial reform is in prospect – Article 5 of the German Unification Treaty provides for the merging of Brandenburg and Berlin in one federal *Land*. As we have mentioned, attempts to combine the *Länder* of Berlin and Brandenburg, with the support of the governments and parliaments of both *Länder*, were rejected in a referendum. Most notable, it was rejected by a majority of the population of the new *Bundesland* Brandenburg. It is clearly open to question whether it will come to pass before the end of the century. At present, especially in the new *Länder*, there is an abundance of difficult and more immediate problems in most spheres of internal German politics. In all probability the task of restructuring the entire federal system will be postponed indefinitely in the face of these current problems.

It is hard to imagine any event in Germany's recent past which has had as much significance for national as well as European history as German unification.

At a time when hardly any politicians or citizens of the *Bundesrepublik* considered 'German reunification', which was laid down in the preamble to the *Grundgesetz* as a prime policy objective, to be a real possibility, the socialist camp broke up at breathtaking speed, and socialism as both an ideology and a political system collapsed throughout Europe and therefore also in the DDR. The opportunity for the reunification of Germany appeared, then, quite unexpectedly; the citizens of the DDR, who had just brought about the surrender of their socialist masters with the words '*Wir sind **das** Volk*', now clearly expressed their political will for a rapid national amalgamation of the *Bundesrepublik* and the DDR with the new slogan '*Wir sind **ein** Volk*'. In the autumn of 1989 and spring of 1990, this political will built up such a momentum that it swept aside all resistance to German unification, but also all in-between measures.

It is important for our theme that in this process there was also a revival of the federal idea in East Germany. The collapse of the centralist SED regime, which had abolished the original federal structure of the DDR in the early 1950s for reasons of political power, brought with it tendencies towards regionalisation on the basis of the old traditional *Länder* in the DDR. Democratisation and federalisation therefore came into being simultaneously as political options for the citizens' movement at the end of the DDR. The re-establishment of the former *Länder* and of communal self-government therefore formed part of the primary political demands of the citizens' movement in East Germany, among others expressed in the 'round table' negotiations. The political parties and democratic groups which were anxious to gain influence over the democratisation process in this first phase, could not ignore demands for the federalisation of the DDR. Thus the first and only freely-elected and democratically legitimised DDR government, which followed the *Volkskammer* elections on 18 May 1990, announced as one of its first aims 'the creation of a federal republic'. Accordingly a 'Ministry for regional and communal affairs' was set up. Several different models of federal structure, with varying numbers of re-founded *Länder*, were discussed in the *Volkskammer*. This discussion was soon caught up in the slipstream of the reunification discussions. In the end that led to a majority view in favour of the re-establishment of the five former *Länder*, Mecklenburg-Vorpommern, Brandenburg, Sachsen, Sachsen-Anhalt and Thüringen (Mecklenburg-West Pomerania, Brandenburg, Saxony, Saxony-Anhalt and Thuringia), in spite of considerable misgivings over the existence and effectiveness of these *Länder*. On the basis of these discussions, the DDR Volkskammer formulated a law (*Ländereinführungsgesetz*) which set the date of 14 October 1990 for the formation of five *Länder* and laid down the rules for every essential aspect of the new federal structure of the DDR. These included the responsibilities of the *Länder*, the distribution of legislative and financial responsibilities, etc., and indeed the relationship to the *Grundgesetz* of the *Bundesrepublik*, in order to facilitate subsequent unification.

This process of refederalisation of the DDR was, however, overtaken by the speed of German unification. The *Ländereinführungsgesetz* should have come into force in the DDR on 14 October; yet on 3 October the unification of Germany through the accession of the DDR *Länder* to the *Bundesrepublik* was completed according to Article 23 of the *Grundgesetz*. The governments of the two German states had agreed this in the 'treaty between the *Bundesrepublik Deutschland* and the *Deutsche Demokratische Republik* on the restoration of German unity' (Unification Treaty of 31 August 1990). As early as 1 July 1990, the treaty establishing economic, monetary, and social union, by which the German Mark (DM) became the official currency of the DDR, had come into force.

In order that the federal structure of the *Bundesrepublik* should not be put at risk by the speed of this reunification process, the West German *Länder* had proposed, on 5 July 1990, a number of parameters (*Eckpunkte*) for federalism in a united Germany. Their intention was to underline emphatically the federal character of reunited Germany, while at the same time speaking up for the extension of federalism, and particularly for a strengthening of the position of the *Länder* in the constitutional system.

The new *Länder* could only be constituted properly once unification had taken place, following the *Landtag* elections on 14 October 1990 and the formation of the *Länder* parliaments, which at the same time had to function as constituent assemblies. These imposed provisional constitutions on the *Länder* and decided on the capital of each *Land*. It was only after the formation of the new *Länder* governments that the *Bundesrat*, representing the *Länder*, was able to assemble. This it did on 9 November 1990, with the representatives of 16 rather than the previous 11 *Land* governments. Thus the refederalisation of the former DDR and the integration of the five new *Länder* into the federative system of the *Bundesrepublik* was completed. The problem now is to breathe life into this newly-integrated structure, and to make it possible for the new *Länder* to take their rightful constitutional place within the federal constitutional system of the *Bundesrepublik*. This may well not be easy; the generally impoverished condition of the new *Länder* means that, for the foreseeable future, they will be in an extremely weak position and in need of help, above all in the form of financial support on the part of the *Bund* and the old *Bundesländer*. It is doubtful, then, at the very least, whether the increase in the number of *Länder* through the unification process can produce a qualitative strengthening of federalism. The dependency of the new *Länder* is enormous, and is bound to last for a long time.

Current problems of German Federalism

Recent significant problems affecting German federalism stem from two different political developments – first from the process of German unification, and second from the process of European integration.

Both problem areas represent considerable challenges to traditional German federalism. There is no prospect of a straightforward, rapid and satisfactory solution to either, especially since the political system in the *Bundesrepublik* is currently undergoing a difficult endurance test, perhaps even the most difficult in its history.

We have seen that German unification followed the accession of the five *Länder* of the former DDR in accordance with Article 23, which made possible the accession of 'other parts of Germany' to the *Bundesrepublik*. Through this form of accession, German unification became a problem of German federalism, to the extent that the new *Länder* were to be included in the German federal system. In terms of constitutional law, this means giving the new *Länder* the same status and the same rights as those possessed by the old *Länder* since the founding of the *Bundesrepublik*, including the revival of previous *Land* traditions, the forming of suitable *Land* constitutions, political systems and parties, etc.

For the new *Länder* to achieve formal legal parity with the old *Länder* they require, along with their formation as political entities with constitutions, parliaments, governments, etc., equal status within the framework of the *Bund*. This is necessary in view of their involvement with the federal state financial constitution, with the distribution of legislative responsibilities both in the *Bundesrat* and as *Länder* representatives at the federal level.

Bringing the new *Länder* into line with the old, as political entities with democratic constitutions and institutions, with their democratic legitimisation through elections and with their inclusion as members with equal rights in the *Bundesrat*, created no major problems. However, the alignment of the new *Länder* with the old in the context of the financial system is causing considerable difficulties. The problem is that all five new *Länder* are relatively poor and weak and will be seriously in need of substantial and guaranteed long-term support from the *Bund* and the old *Länder* after 40 years of a socialist regime. In view of the present serious difference of level, it was not possible for them to be admitted immediately into the financial equalisation system of the *Bundesrepublik*. Instead, the unification treaty of 1990 provided for the new *Länder* to be financially supported until 1994 by the 'German Unity Fund' (*Fonds Deutsche Einheit*) set up specifically for that purpose. In the period 1990-94, a total of 115 billion DM was made available by the *Bundesrepublik* (*Bund* and old *Länder*), with little visible sign of success to date. That is why further efforts are called for, leading amongst other things to the so-called 'solidarity pact' of *Bund* and *Länder*. In years to come a gigantic redistribution from West to East will be required, in order to re-establish equal living standards in Germany as a whole. To start with, this means a fall in the living standards and prosperity of the old *Länder*. The crux of these efforts is the economic reconstruction of East Germany, whose industry faces the most severe difficulties in all areas because of mismanagement during the DDR era. The economic problems of unification have been grossly underestimated by the federal government. This has led to many misjudgements and the poor planning of reconstruction policies, resulting in the collapse of many industries and growing unemployment in the new *Länder*. The consequences of these problems for German political culture are incalculable. In the new *Länder* particularly, once the initial euphoria of German unification had worn off, disillusion has been widespread. It is unfortunate that the construction of a new democratic way of life has coincided with an economic recession and a crisis in living standards. It will not be easy to win back the trust in democratic politics and a democratic political system which has already been lost. These negative trends also have a negative effect on the acceptance of federalism and its achievements. In recent years a special tax, called the 'solidarity tax', was introduced to enable more money to be devoted to construction in the new *Länder*. However, this 'solidarity tax' is to be reduced from 7 per cent to 5 per cent in 1998.

The process of European integration also constitutes a serious problem for German federalism. The *Länder* are under threat, in terms of their decision-making power, from the progressive transfer of responsibilities away from the federal government towards Brussels. Their participation in political decisions has been severely reduced also at the federal level, their responsibilities eroded and their substance threatened. There is a danger that matters of particular concern to the *Länder* will remain unconsidered. In this situation, which has led to considerable tension between *Bund* and *Länder*, the *Länder* have pursued three means of defence against the obvious danger to their political existence – first, they have established their presence at the centre of European Community decision-making by setting up special offices in Brussels and, as a form of *Länder* lobby group, attempted to gain direct influence there on European policy-making; second, they have sought to improve their position in the European decision-making process in relation to the ratification of the Maastricht Treaty, an attempt which is reflected within the *Bundesrepublik* in the revision of Article 23 of the Constitution. Here it is expressly stated that 'The *Bundestag* and the *Länder*, through the *Bundesrat*, shall participate in the affairs of the European Union. The Federal Government shall keep the *Bundestag* and the *Bundesrat* informed, comprehensively and at the eatrliest possible time'. It is later stated 'the *Bundesrat* shall participate in the decision-making of the Federation insofar as it would have been competent to do so in a comparable internal action, or insofar as the *Länder* would have been competent to do so internally'; and a third possibility, or opportunity, to safeguard and perhaps even to strengthen the position of German federalism within the framework of European integration has arisen recently through the increased coordination of regionalist efforts at a European level, efforts in which the *Länder*, through their initiatives, have played a leading part. For a long time the *Länder* scorned the prospect of cooperating with regionalist groupings and organisations in other countries because they did not view them as having genuine constitutional status. Latterly, however, they have overcome these shortsighted inhibitions. They are cooperating with other European regions of varying constitutional status and trying to institutionalise, in an official 'Committee of the Regions', the specific needs and interests of each European region. Behind this lies the awareness that the traditional nation states are often unsuited, as members of the European Community, to an appreciation of regional interests, that these regional problems require the representation of their own interests through the regions themselves. The sociologist Daniel Bell has found a convenient formula to describe this phenomenon – the traditional nation state is now too small for the big problems and too big for the small problems. That is to say that he is proposing a reinforcement of the transnational political level at the same time as the construction of a regional level – and both of these, in the final analysis, would be to the cost of the nation state. Within this context, the *Länder* of the *Bundesrepublik* can play a particularly meaningful exemplary role, as developed political entities with constitutional status. This role became even more significant when in 1995 Austria, with its similarly structured federalist *Länder*, was admitted to membership of the EC. The European Union (as it is now), which is moving towards political union, will probably only be able to turn this political union into a reality if it is a federal one, in which the regions are granted an independent and responsible role at a lower level. The introduction of the subsidiarity principle in the Maastricht Treaty also points in this direction. Therefore, in this newest treaty on European integration, a socio-philosophical principle has been introduced that stems recognisably from the social teachings of the Catholic Church. It makes the prescriptive statement, which is as it were grounded in natural law, that the lower and smaller unit takes precedence in being allowed to function according to its natural functional capacity. As a result, these smaller or lower units, the regions or *Länder* in the federally structured entity of Europe, are guaranteed preferential freedom of manoeuvre when it comes to determining what they are. This policy is reinforced by the Amsterdam Treaty, concluded in 1997 and awaiting ratification by national parliaments.

Conclusion

In the mid-1950s, it was not uncommon to hear calls for the abolition of a federalism seen as obsolete, faced with the tremendous processes of industrial and economic development of the old *Bundesrepublik*. On the one hand federalism was seen as a hindrance to prosperity and progress, on the other as an unnecessary and expensive antiquity, with its multiplication of state institutions (parliaments, governments etc.) at the level of the *Länder*.

This sort of reaction is no longer heard. In recent years there has been a fundamental change to the long-held belief in the supreme value of state unity as opposed to traditional and regional variety. Paradoxically, in the age of trans- and supra-national cooperation and integration, the level of regions and *Länder* has gained in significance throughout Europe. The view is widely accepted that those nation states organised along the most centralist lines have become doubly problematical today – they appear too small to deal with the big problems, and too big to deal with the small ones. In the first case, it is trans- and supra-national integration which can provide a response, in the second, the formation of regions and (trans-national) regional cooperative entities. The attempt at integration embodied in the Maastricht Treaty has taken this tendency into account through the introduction of the concept of subsidiarity. In the preamble to the treaty, the European Community Member States declare their common determination 'to continue the process of creating an ever-closer union among the peoples of Europe, where, in keeping with the principle of subsidiarity, decisions are taken as closely as possible to the citizens'.

This European development is naturally reflected at the level of the individual nation state. As far as the *Bundesrepublik* is concerned, this means that the principle of federalism enshrined in the German constitution can no longer be seen as politically outdated and irksome. On the contrary, both with regard to the further development of Europe as well as, most especially, to the process of the completion of German unification, the federal structure of the *Bundesrepublik* occupies a key position, and one which, with regard to other European states organised until now along centralist lines (e.g. Great Britain, Italy, France), could exert an important exemplary influence. This does not mean that the obvious problem areas of current German federalism can be overlooked; on the contrary, present discussions in the *Bundesrepublik* on the revision of the constitution contain, amongst other things, proposals for the reform of the federal system. For example, the revision of Article 23 of the *Grundgesetz* indicates that new rules had to apply to the *Länder's* right to participate in European decision-making. The capacity for reform of the federal system in the *Bundesrepublik* will have a significant impact on its future. Rapid political change, both internal and external, is forcing federalism continually to adapt, through the modification of structures and procedures. With this vitality, the federalist principle can represent in the future, in cooperation with the other constitutional principles, an important determining factor in the political system of the *Bundesrepublik*. In today's world, and in spite of all efforts at integration, politico-constitutional developments point in the direction of diversity, and it is this which, to a far greater extent than uniformity, gives free rein to the expression of human experience and needs in all their variety.

References:

Benz, A. (1993), 'Redrawing the Map? The Question of Territorial Reform in the Federal Republic', in Jeffery, C. & R. Sturm (eds), *Federalism, Unification and European Integration*, London, Frank Cass, pp. 38-57.

Currie, D. P. (1994), *The Constitution of the Federal Republic of Germany*, Chicago & London, University of Chicago Press. [An English-language text of the Basic Law (*Grundgesetz*) of 23 May 1949, amended 1 December 1993, appears as an Appendix to this volume, pp. 343-412.]

Exler, U. (1993), 'Financing German Federalism: Problems of Financial Equalisation in the Unification Process', in Jeffery, C. & R. Sturm (eds), *Federalism, Unification and European Integration*, London, Frank Cass, pp. 22-37.

Gunlicks, A. B. (1995), 'The "Old" and the "New" Federalism', in Merkl, P. H. (ed.), *The Federal Republic of Germany at Forty-Five*, London Macmillan, pp. 219-242.

Jeffery, C. (1998), 'German Federalism in the 1990s: On the Road to a "Divided Polity"?', in Larres, K. (ed.), *Germany since Unification*, London, Macmillan, pp. 107-128.

Kilper, H. & R. Lhotta (1996), *Föderalismus in der Bundesrepublik Deutschland: eine Einführung*, Opladen, Leske & Budrich.

Federalism in Austria

Josef Honauer

Introduction

With an area of 83,853 square kilometres and a population of slightly less than eight million, Austria is one of the smaller European countries. But this has not always been the case. Before the First World War, the Austro-Hungarian Empire was a leading power in Europe. After the old empire disintegrated, Austria became a federal republic. In 1938 it disappeared completely from the map during the Third Reich, only to be re-established under the guidance of the occupying allied forces. Partly as a consequence of its eventful past and geographic conditions, the country's nine federal provinces (*Bundesländer*) – Burgenland, Carinthia, Lower Austria, Salzburg, Styria, the Tyrol, Upper Austria, Vienna and Vorarlberg – retain their local characteristics. Due to the country's small size every *Bundesland*, the capital Vienna excepted, borders on one of Austria's eight neighbouring countries. Ties from earlier epochs in history are still apparent, and the former Iron Curtain and its abolition continue to have a massive influence on politics and society as a whole.

On the basis of a constitution which is committed to federalism, the provinces enjoy a certain amount of autonomy, though not as much as the German *Länder* (Nick, 1992, p. 60). A significant imbalance is caused by Vienna, located in the East, because of its political and geographical position and its size, which, in relation to Austria as a country, is far too big. Having grown rapidly in the 19th century, it was once the multicultural centre of the Austro-Hungarian Empire's 53 million people (Weinzierl, 1995, p. 200). Since then Vienna has remained 'rot', i.e. socialist, as has the Burgenland, ruled by the Social Democrats (*SPÖ – Sozialdemokratische Partei Österreichs*) in contrast to all other *Bundesländer*, which are 'schwarz', i.e. conservative, governed by the People's Party (*ÖVP – Österreichische Volkspartei*).

Politics in Austria has long been dominated by the two largest parties, the *SPÖ* and the *ÖVP*. These were founded in the 19th century, earlier than the country itself, and dominated the political landscape not only during the First Republic, when fighting between the two factions culminated in civil war, but also in the Second Republic where the lessons from the First had been learned and the leaders of both parties agreed to cooperate. Austrian corporatism, the *Sozialpartnerschaft*, has long dominated interior affairs, and has helped ensure social stability, with very few industrial disputes (Markovits, 1996, p. 16). Only recently has this policy been questioned by the leader of the younger and extreme right-wing party, the *FPÖ* (*Freiheitliche Partei Österreichs*), which has increased its appeal by means of xenophobic criticism of immigrants and

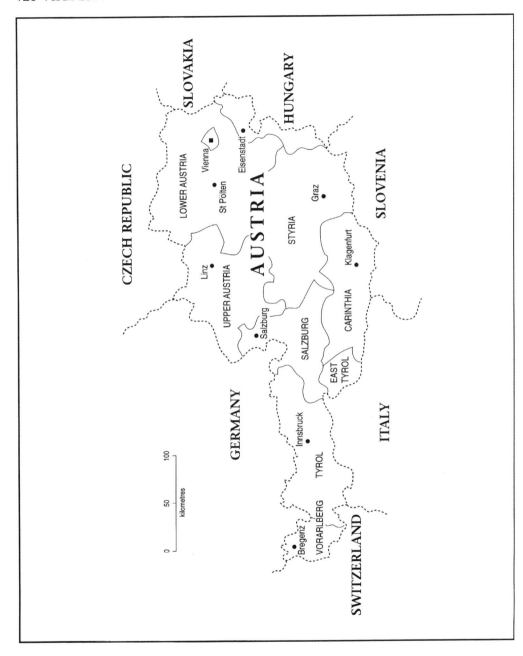

refugees. In the mid-1980s the green party, *die Grünen*, and in 1992 the liberal party, *Liberales Forum*, also secured seats in the *Nationalrat*, Austria's parliament, but the two main coalition parties still hold two thirds of the seats by a very narrow majority.

The influence of a few major agglomerations such as Vienna, Graz, Linz, Innsbruck and Salzburg notwithstanding, infrastructure policy continues to be dictated by the country's geography. Transport and transit especially seem to present a contradiction in foreign as well as internal policies. With its central European situation, Austria forms a crossroads between North and South, East and West. Its negotiating position over transit within the European Union has, however, been considerably undermined since joining the Union in 1995, and the Tyrol suffers especially from the immense impact of thousands of lorries travelling through this corridor connecting Italy to Germany.

Origins and History:

Before 1918

Austria has a long and complex history which dates back to the Middle Ages. A first mention of '*Ostarichi*' in 996 referred broadly to the valley of the River Danube which crosses Upper and Lower Austria. It was founded by the Holy Roman Emperor Otto III to defend western Europe against peoples from the East (Scheuch, 1994, p. 22). First the Babenberger and later the Hapsburg ruling dynasties were able to increase their power, importance and hence territory, largely through arranged marriages rather than through military conquest.

Ostarichi grew gradually, first towards the South absorbing Styria and Carinthia, then to the West annexing the Tyrol, Salzburg and later Vorarlberg and extensively to the East, dominating non-German-speaking areas. Vienna became the cultural and political centre of this huge monarchy ruling over more than 15 nationalities (Hamann, 1996, p. 129). Strong linguistic links between Bavaria and Austria obscure a significant admixture of Czech, Slovenian, Croat, Slovak, Hungarian, Italian, Celtic, Roman and even Illyrian heritage (Zöllner, 1995, p. 30 ff).

After 1918: moulded by the old empire.

'Austria is what's left over.'

Georges Clémenceau (Weber, 1995, p. 69).

The structure of Austria as we know it today consists broadly of the German-speaking parts of the old empire, except South Tyrol which was ceded to Italy, and the Sudetenland in the North, which was lost to the new Czechoslovakia. Between 1919 and 1921, referenda were held in border areas and a number of communities were gained from or lost to Yugoslavia. The most important event in this period was the referendum in the far East region, former West Hungary, where the population voted in favour of Austria, amongst them Croatians, Romanies, Sinti and Hungarians, thus forming the youngest province, the Burgenland.

Towards the end of the 1920s and during the 1930s, recession and record unemployment exhausted both politicians and citizens, and democracy was eliminated in 1933. Several years of austrofascism led directly to the *Anschluß* in 1938, with its disastrous consequences.

After 1945: moulded by war and occupation.

The post-war occupation of Austria by American, British, French and Soviet forces is another significant factor in regional development since the different authorities had no concerted policy and treated their occupied territories differently. While the western forces, mainly the United States, granted financial help in the form of the European Recovery Program (ERP) the Soviet Union which freed and occupied all of Lower Austria, the Burgenland, and the northern part of Upper Austria, the *Mühlviertel*, took a different view and dismantled industrial plants as well as confiscating transport systems and cargo ships (Weber, 1995, p. 73). Similar incidents took place, though not as extensively, in the French-occupied Tyrol and Vorarlberg. The ERP fund was invested 81 per cent in the western provinces, while the traditional industrial locations in Vienna and Lower Austria and the rural Burgenland had to share the rest. As the main period of reconstruction of a largely destroyed Austria coincided with the allied occupation, those regions under Soviet influence did not have the same opportunities as the American-controlled districts, which were able to achieve enormous economic growth, especially in the agricultural sector. American intervention even prevented the new Austrian government from selling production plants from the steel industry established under Nazi-Germany in Linz, since the Americans saw it has having been designed for a Greater Germany and therefore as too big for Austria. After the *Hermann-Göring-Werke* were nationalized and renamed *VÖEST* they became one of the biggest employers in Austria and were the foundation of healthy economic growth in Upper Austria. In addition, the degree of destruction by allied bombing was twice as high in the East as in the West. While in the east nearly 60 per cent of the rail network was put out of action, the west suffered a destruction of less than 30 per cent (Weber, 1995, p. 74). Although the existence of four different zones was an obstacle to both trade and traffic, the major impediments to industrial development were to be found in the shortage of energy, raw materials and items of capital expenditure. The western provinces were better provided with food, energy and a much more rapid re-establishment of the transport systems. They therefore underwent a faster redevelopment in comparison with the eastern provinces in the Soviet zone. The Marshall Plan (economic and technical aid provided by the United States for non-Communist European countries after the Second World War) enforced the regional shift of focus towards the West. The underdevelopment of East Austria, left behind during the reconstruction period, thus reinforced a regional distinction which had other causes. First, Nazi Germany had invested in the western provinces rather than in the Vienna-dominated East. After 1945, moreover, the closeness to the Iron Curtain and, until 1955, the psychological effect of the presence of the Red Army, were contributing factors.

Constitution and Federalism

Every state-building process, even seen as a continuing phenomenon, has to face two challenges – outward penetration from a centre into its provinces, and integration from the periphery towards a centre. The most important prerequisite for this is the pre-existence of such a centre (Marko, 1992, p. 731). Vienna, as the focal point for finance, trade and administration, has long fulfilled this requirement. According to all comparative social studies, Austria, unlike her German-speaking siblings, is clearly a monocephalic country, centre-orientated in comparison with the polycephalic structure of Germany and Switzerland. Even with a population in decline since the First World War, the structural tensions between an economically dominant capital and its sluggish provinces grew because of the political tensions between a socialist-dominated Vienna and the more conservative *Länder*. Looked at from this perspective, the austrofascistic *Ständestaat* (1933-38) can be interpreted as a victory of the country over its capital, which was referred to as the *Wasserkopf*, the hydrocephalus (Scheuch, 1994, p. 170). Anti-Vienna tendencies were enforced during the Third Reich, when Vienna was downgraded to the status of a German provincial city. After 1945 the capital, with its quadripartite occupation and surrounding Soviet zone, was isolated from western influences. Tensions between the centre and its periphery continued, but in comparison to the pre-war period these lacked the ideologico-political dimension and were therefore peaceful. A well-established and organised civil service meant that the Soviet strategy of penetration could not be thrown off course, but tightened the bond between Vienna and the federal provinces. At an earlier stage, some of the latter had even wanted to secede, and a failed attempt by the Tyrolese to create a republic of their own in 1919 was followed by an equally abortive referendum in Vorarlberg, whose inhabitants wanted to merge with Switzerland. The Swiss, despite evident ethnic links (the easternmost province, Vorarlberg, is inhabited by people of Swabian stock, historically known as the Alimanni and who speak a dialect which bears similarities to Swabian and Swiss German) turned them down. After 1945, none of the federal provinces tried to break away, and despite evident tensions between the eastern and western provinces, integration itself has not been questioned since then.

Democratic Institutions

'Austria is a democratic republic. Its law emanates from the people.'
(Austrian constitution, Article 1).

The Austrian parliament has two houses – the *Nationalrat*, with its 183 directly elected members, and the *Bundesrat* with 63 members who are delegated by the *Landtage*, the nine provincial assemblies. Draft laws supported by a minimum of 100,000 qualified voters can be put to the *Nationalrat* for ratification. The president of the Republic, whose political role is largely symbolic, is elected by direct popular vote for a six-year term, renewable once. This head of state appoints a Federal Chancellor, normally the chairperson of the strongest political party, who in turn appoints a cabinet. *Landtage* consist of only one chamber in comparison with German *Länder*, which have two, and they cannot be replaced by people's assemblies, as in Swiss cantons. They are elected by the residents of the federal province. As long as the rules of local elections do not conflict with federal ones, *Länder* are free to change election laws, but they are obliged to retain a system of proportional representation.

Landtage enjoy a monopoly of legislation (*Gesetzgebungsmonopol*) and instruments of direct democracy such as referenda, which would subsume this monopoly, are therefore against the constitution (Marko, 1992, p. 729). The contribution of the *Länder* to the federal legislative process has limited political impact since the *Bundesrat* only has the power to veto a bill once. The governments of the *Bundesländer* consist of a Minister President (head of a province, or *Landeshauptmann*), his deputy and the requisite number of members of the government of a province (*Landesräte*). The governments of the *Länder* are subject to instructions by the federal government but the *Landeshauptmann* himself has the authority to issue directives. The *Landtage* are elected every five years, or, uniquely in the case of Upper Austria, every six years, and the *Landeshauptmann* is appointed by the members of the *Landtag*. In 1996 the constitution of Carinthia was changed to create a more effective procedure for the election of the *Landeshauptmann*. Since then it has been impossible to block his election for weeks on end – as happened in 1994, when members of the *FPÖ* refused to vote for a social democrat. Now in the third ballot only a simple majority is required. Every *Landesregierung* has to be a so-called 'concentration government' (*Konzentrationsregierung*), which means that every party receiving more than a certain percentage of votes has to be represented in it. Sometimes the leading parties work so closely together that they even intervene in each others' party business. For example, when the newly designated Prime Minister (*Kanzler*), Viktor Klima, wanted Burgenland's *Landeshauptmann* to become his chancellor (*Finanzminister*) in 1997, the coalition partner opposed this idea because it did not want to have to do without him (*Standard*, 12 January 1997). Following recent discussions on the reform of the federal state, proposals have been made to replace 'concentration government' by majority government. The effect of this would be to change radically the political function of the *Landtage*, with the intention of guaranteeing greater political independence for the *Länder*. At the moment, political parties in the *Landtage* which hold government posts but do not belong to the party of the *Landeshauptmann*, find themselves in a rather delicate position – on the one hand they share government duties and responsibilities but at the same time they form the opposition. In contrast to the *Nationalrat* there is no pressure from whips to toe the party line. Not only do the provincial governments (*Landesregierungen*) consist of members nominated according to a system of proportional representation, but all institutions to be appointed by the *Landtag* do so too. This includes education boards, public transport committees, public media facilities, and the tentacles of this *Proporz* (collective power shared by the two main parties to the exclusion of all minority parties) even extends into public radio and TV as well as to the nomination of managers within nationalised industries and other businesses owned by the public sector, such as housing associations. All these areas are not only important in terms of political influence but also as profitable sources of income.

The lessons learned in the inter-war period with its constant threat of insecurity which led to civil war, economic, political and social, in 1934, has predisposed all players in the game to cooperate closely. All government committees and sub-committees consist of all parties represented in the *Landtag*. The standing orders of procedure in Carinthia and Lower Austria even provide for the chairpersons of committees to be elected by the *Landtag* by proportional representation, but not according to how well-qualified an individual may be in a particular field. In addition, members of *Landtage*, even chairpersons of committees, can be replaced at any time by the leadership of their party. It should be noted that, in practice, and because of the majority enjoyed by the two largest parties, the number of committee members can be changed and the *Proporz* distorted in favour of these two parties, especially when they want to prevent smaller parties from gaining seats and having a say. This is what happened in 1988, when the *FPÖ* managed to gain representation in the *Landtag* of Lower Austria for the first time, but without wielding any real power on committees. On the supervisory committee only the party of the *Landeshauptmann* is represented, a peculiar state of affairs given the predominance of proportional representation elsewhere.

Where legislation is concerned, the *Landtage* are influenced not only by the political parties but also by pressure groups (*Verbände*) representing the three principal estates – business and industry, organised labour and the farming community. Permanently represented at government level through their respective federal chambers, their delegates occupy seats in the *Landtage* and in the local governments of the *Länder*. Faced with the increasing influence of the *Verbände* on legislation, Carinthia's *Landesregierung* took a step forward and decided that a chair in a pressure group is not consistent with a seat in the local government. One can doubt whether this reflects an appropriate balance of power, since a political position in the legislature and a leading post in interest groups, such as chambers of commerce, should not be mutually exclusive.

Where the balance of interests between provinces (*Länder*) and country (*Staat*) is concerned, there is a general clause in the constitution which grants overriding authority to the federal provinces. However, there are so many exceptions to the rule that only the following areas actually remain within the competence of the *Länder* – planning regulations and development, nature preservation, hunting and fishing, waste removal, pre-school education and day nurseries. The *Länder* are also responsible for the enforcement of a range of federal laws relating to social welfare, hospitals, spa resorts, electricity, land reform and primary and secondary schools in the public education sector. All other legislative power resides with the state government in Vienna.

Apart from the legislative process itself, the *Landtag*, in theory, has an important role to play in monitoring and controlling government decisions. It has to be said that, especially in this respect, none of the *Länder* has mechanisms in place which enable it to do justice to a modern multiparty parliamentary system of government. Most of the monitoring functions are geared entirely to the party which holds the majority in the house, so that, for example, a majority vote is required to establish investigating commissions. The growth in numbers of these investigating commissions in different

Landtage is a reflection both of the effects of investigative journalism and of the existence of a need for such an instrument. In Vorarlberg the reform of this section of the constitution now enables an investigating commission to be set up with the support of just one third of the members of the local parliament. Furthermore, the importance of financial as well as political control mechanisms can be seen in the attempts which have been made to achieve a greater degree of transparency in financial matters. In Styria in 1982, for example, an independent audit office was established as a first step towards an extensive reform of the provincial constitution. Salzburg followed this example in 1984 and since then in many other federal provinces this issue has had a high priority. A major innovation in Styria has been the institutionalised combination of control measures as a means of direct democracy. This new law provides for investigations by the local audit office at the request of two per cent of the local population. Especially in this area, the opposition parties are attempting to improve their image as the controlling party in reclaiming the chair of the audit committee. Legal control through the federal government in Vienna is the only possible way to keep a check on a *Landeshauptmann*. For example, in 1985 the *Landeshauptmann* of Salzburg faced a charge from the Secretary of State for Labour and Social Affairs because he allowed businesses to stay open on 8 December, which is a national holiday when all shops are supposed to be closed.

The principles of majority-based decision-making as described above have been stable for a very long time. This, combined with the role of concentrated local governments, has ensured a relatively calm and cooperative atmosphere, but one which is, at the same time, static and less than dynamic. Since a minimum of between 4 per cent and 6 per cent of votes is required to guarantee a seat in the *Landtag*, small opposition parties had no significant role until the 1980s, when the era of the mass parties started to decline. It is these small parties which are now filling the vacuum left by the traditional opposition in the *Landtage*.

Between the democratic institutions of the *Länder* and the equally democratic institutions of the communities (*Gemeinden*) there is no intermediate level of democracy. Regional authorities as such exist only at the level of public administration. An administrative district (*politischer Bezirk*) consists of several communities grouped together, but with no democratic accountability. At the head of each administrative district is a *Bezirkshauptmann* who is appointed by the *Landeshauptmann* and whose function is that of a civil servant subject to instructions. There is no parliamentary representation at this level, with two exceptions – first, some communities, the larger towns, are so-called 'cities with their own *statutes*' (*Städte mit eigenem Statut*). In this instance the responsibilities of community and district authorities overlap, so that the position of the *Bezirkshauptmann* is replaced by that of a *Magistratsdirektor*, subject to instructions from the mayor, who is the elected head of the city council. Second, Vienna, which is *Land* and community in one, is divided into districts. These are not administrative districts but parliamentary units similar to communities or city councils. Vienna's role as the federal capital (*Bundeshauptstadt*), federal province (*Land*) and community (*Gemeinde*) at one and the same time was further complicated until 1997, when the *Landesregierung* of Lower

Austria was transferred from Vienna to the new provincial capital (*Landeshaupstadt*) of St. Pölten.

It is the community (*Gemeinde*) which constitutes the base level of Austria's political structure – the lowest level at which democratic decision-making takes place. As a consequence of the social structure of the Austrian party system, the social-democratic *SPÖ* has traditionally held the majority in larger towns, while the conservative *ÖVP* has held sway in the smaller communities. Since the 1970s this division has become less clearly defined because the SPÖ has been able to take advantage of a general trend in society to claim more seats within smaller communities (the *ÖVP* has managed in coalition with the *FPÖ* to defeat the *SPÖ* in a few councils). Despite these conflicting tendencies, politics in the big cities and in urbanised environments is still generally the domain of the social-democrats, whereas in rural areas the conservatives wield the greatest influence (Nick, 1993, p.62). Basically, communities are structured in parliamentary fashion and the parish councils (*Gemeinderat*) are elected directly by the residents. In most communities the Mayor is installed by the local council but in the 1990s more and more *Länder* are beginning to have a direct vote on the mayor independently from the council.

National Identity – Regional Identity

During the 19th century, when the countries of Western Europe developed into nations, Austria exemplified the outer limits of ethnic, national and religious diversity. The Hapsburg's success in establishing a well-organised state was not matched by their ability to build a nation (Hanisch, 1992, p. 16). Linguistic frontiers marked the boundaries as different peoples developed into nations, in contradistinction to the multinational Empire. The German-speaking Austrians formed a German nation in opposition to the other ethnic groups and, since the ruling dynasty was German-speaking, they identified themselves with the Empire and formed the 'ruling nation'. During the time of the monarchy Austrians developed a double identity – one which was strongly German, founded in language, education, and culture, and a weaker, Austrian one which was symbolized by the Empire. But how German is Austria? The desire of the Austrians to be Germanic culminated in the *Anschluß*, a form of marriage leading to massive disillusionment as it proved to involve unexpected domination by 'Prussians'. It was only after 1955 that a process of Austrian nation-building began in earnest, closely connected to the economic miracle of the post-war years (*Wirtschaftswunder*). All studies now indicate that the population has an increasingly strong sense of Austria's nationhood. Fifty years after the Second World War and 40 years after the reacquisition of its political sovereignty and independence, 85 per cent of Austrians questioned think that Austria is, in fact, a nation (in contrast with a figure of 47 per cent in 1964), while 9 per cent believe that nationhood is in the process of being created (1964: 23 per cent) and a mere 4 per cent (1964: 15 per cent) think that Austria has no claim to that status (Haller, 1996, p.66). All the indications are, therefore, that Austrians do indeed have an Austrian identity. A more precise picture emerges from a study of intra-Austrian diversity.

Identification with regional-territorial units follows a pattern of concentric circles.

The innermost circle corresponds to the affiliation to a community or, within towns and cities, to a neighbourhood. In these units individuals occupy a position in a close network of friends, family and neigbours with a relatively low degree of social distinction. To explain this level of affiliation sociologists use the term 'place-attachment' (*Ortsverbundenheit*), one of Austria's most highly-valued characteristics. In the opinion of the German magazine *Wirtschaftswoche*, Austrian workers are industrious, intelligent and take a pride in their work, but are not mobile enough. Only one in five would be ready to move to a different city or province, even if failure to do so would lead to unemployment (Haller, 1996, p. 392). The next highest level, the second circle, is the region, which has a different definition in every country and, even within Austria, in every single federal province. Regions do not always correspond to administrative districts, but increasingly discussion is focusing on their role, because of their geographical, socio-cultural and political importance. It is administrative districts that form the basis for the European Union-promoted 'NUTS' (Nomenclature des unités territoriales statistiques), which themselves focus on geographic-functional conditions. There are, however, historically established regions which do not necessarily correspond to NUTS, just as there are political districts with which people identify themselves. Lower and Upper Austria are traditionally divided into quarters (*Wald-, Wein-, Most-, Industrieviertel; Mühl-, Inn-, Hausruck- und Traunviertel*). The *Wald-* and *Mühlviertel*, both situated north of the river Danube, suffered a delayed economic development because they had been occupied by the Red Army. Later, due to the Iron Curtain, no transnational exchange with neighbouring countries was possible and the whole region along this dead border remained less industrialized, with a significant migration to Linz and Vienna. In Salzburg the regions were known as *Gaue* long before this term was used by the Nazis to rename certain places (*Pinz-, Lung-, Pongau* etc.). These regions have grown sociographically, historically and culturally over long periods. Economically interwoven, they form more or less homogeneous areas which overlap with other *Länder*. The best example is the *Salzkammergut* which stretches over three *Länder* – Upper Austria, Salzburg and Styria.

Situated at an intermediate level between the region and the whole country is the federal province, which of course plays a key part in the process of identification and allegiance. How does the relation between local, regional and national identity work? Anyone who feels a strong sense of belonging to his or her community, will normally also identify strongly with the *Land* and the nation. There is, however, an exception to this generalised sense of allegiance – the Viennese identify themselves less with their federal province than do the inhabitants of other provinces. This might have to do with the fact that the level of education is higher here than elsewhere and that there is a considerable migration from the provinces to the capital. Without the latter, the population of Vienna would be in sharp decline.

One interesting aspect to be considered is the fact that there is an equally pronounced sense of belonging felt by the citizens of the remaining eight federal provinces. There is no *Land* which can claim an exceptionally high degree of pride or identity – not even in the provinces which have traditionally been thought to inspire such allegiance, such as the Tyrol or Styria. The data proves that an average of 90 per cent of the population are proud to be citizens of their *Land* (Haller, 1995, p.401).

This evidence is of some significance in a discussion of the role of the *Bundesrat*, the second house of parliament, whose members are delegates from the *Länder* and which is virtually powerless. It would seem, therefore, that the sense of allegiance and degree of pride felt by citizens for their *Land* is not reflected by representation in the federal parliament. It is issues such as this which appear likely to come to the fore as the delicate balance between national and regional identity is brought into sharper focus by burgeoning European integration.

References

Haller, M. (1996), *Identität und Nationalstolz der Österreicher*, Vienna, Böhlau.

Hamann, B. (1996), *Hitlers Wien. Lehrjahre eines Diktators*, Munich, Piper, 1996.

Hanisch, E. (1992), 'Kontinuitäten und Brüche: Die innere Geschichte', in Dachs, H. *et al* (eds), *Handbuch des politischen Systems Österreichs*, Vienna, Manz.

Marko, J. (1992), 'Die Verfassungssysteme der Bundesländer: Institutionen und Verfahren repräsentativer und direkter Demokratie', in Dachs, H. *et al* (eds), *Handbuch des politischen Systems Österreichs*, Vienna, Manz.

Markovits, A. S. (1996), 'Austrian corporatism in comparative perspective', in Bischof, G. & A. Pelinka (eds), *Austro-Corporatism, past, present, future*, (Contemporary Austrian Studies, vol. 4), New Brunswick & London, Transaction Publishers pp. 5-20.

Nick, R. & A. Penlinka (1992), *Österreichs politische Landschaft*, Innsbruck, Haymon.

Scheuch, M. (1995), *Österreich, Provinz Weltreich, Republik*, Vienna, Brandstätter.

Statistisches Jahrbuch für die Republik Österreich (1996), Vienna, Österreichisches Statistisches Zentralamt.

Tálos, E. *et al* (eds) (1995), *Handbuch des politischen Systems Österreichs: Erste Republik 1918-1933*, Vienna, Manz.

Weber, F. (1995), 'Wiederaufbau zwischen Ost und West', in Sieder, R., H. Steinert & E. Tálos (eds), *Österreich 1945-1955: Gesellschaft, Politik, Kultur*, Vienna, Verl. f. Gesellschaftskritik, pp. 68-79.

Weinzierl, E. (1955), 'Zeitgeschichte im Überblick', in Dusek, Pelinka, Weinzierl, *Zeitgeschichte im Aufriß*, 4th edition, Vienna, Jugend & Volk.

Zöllner, E. & T. Schüssel (1955), *Das Werden Österreichs*, Vienna, Tosa.

Scandinavian regionalism: the case of Sweden

Lee Miles

Introduction

The Nordic region has often been regarded as a homogenous grouping of relatively small nation states perched on the Northern periphery of Europe. Indeed, it is easy to see why the five states of Denmark, Finland, Iceland, Norway and Sweden have been regarded as a distinct region, given their common Viking heritage, in some cases their shared linguistic and cultural backgrounds, and their relative small population sizes – the largest being Sweden with just under nine million people in the 1990s. Until recently and at various times and in differing forms, these nations have been amalgamated into larger states. Looking back in time, the whole Nordic region has been united, for example during the era of the Kalmar Union (1375 to around 1523) and even as late as the 19th-20th centuries, Norway fell under the Swedish crown (1814-1905) and Iceland was part of the Kingdom of Denmark until 1945. Their commonalities have also been facilitated in the last century by the Nordic nations' long-standing reputations (certainly as regards Scandinavia) of being liberal democratic 'unitary' states, with social democratic inclinations and extensive welfare states. To many political scientists, such as Marquis Childs, these states, correctly or incorrectly, represented the archetypal exponents of a 'third way' between the harsh winds of capitalism and the straight-jacket of communism (Childs, 1980). Assertions of a 'Nordic identity' have been based as much on political, cultural, and economic rationales as geographic, especially since the region is not completely distinct in geographic terms, with for example, Iceland positioned on the Atlantic shelf and Denmark situated on the European continental mainland.

At first glance, then, it would seem that when concepts of regionalism are applied to the Nordic case then differences between inter-state regionalism (where a region is transnational and includes territories of more than one state) and intra-state regionalism (where regionalism is defined in terms of differentiating between regions within defined nation-state boundaries) are often blurred. On the one hand, there have been notable movements advocating 'Scandinavianism' or the promotion of 'Norden' within these countries in the past. The Norden Association, for instance, was formed in 1919 to promote educational and cultural links between them. Furthermore, those desiring the establishment of common, but not necessarily distinct, Nordic approaches between these countries were to find more practical outlets after the creation of key institutional arrangements, such as the Nordic Council (1952) (Wendt, 1979; Thomas, 1996) and notable policy initiatives, for instance, the Nordic Passport Union (1958) and the Nordic Common Labour Market (1954). Indeed, some authors such as Nils Andrén, have gone so far as to claim that the Nordic example developed its own particular type of integration, based on selective cooperation between sovereign states – so-called 'cobweb integration' (Andrén, 1967).

At the same time, there have been equally notable attempts at stressing the differences between these countries within the region, especially since nationalism inevitably becomes interspersed with concepts of regionalism, for example in the promotion of a uniquely Norwegian state identity after the dissolution of the Swede-Norwegian Union in 1905. This is not to say that Nordic forms of nationalism are aggressive, for clearly they are not – war between these countries is now unthinkable to your average Scandinavian – but they have acted as a brake on more ambitious attempts at integration, such as the Scandinavian Defence Alliance (1948-49) and various initiatives aimed at establishing a Nordic Common Market (1947-70). With the exception of Denmark and perhaps Iceland, their relatively large geographical sizes (compared to their small populations) means that there are also substantial regional variations within these states. In the cases of Norway, Sweden and Finland, they include arctic and sub-arctic regions in their northernmost territories including nomadic people (the Samis/Lapps) with differing cultural traditions from those in the more cosmopolitan southern parts of these countries. There are 20,000 Sami living in Sweden out of a total population of 75,000 spanning Finland, Norway, Russia and Sweden, whose main occupation is reindeer husbandry, for which they use a geographical area covering one-third of Sweden's territory. Regional identities can therefore be equally important to discussions of inter and intra-state regionalism when applied to the Scandinavian case. Indeed, the majority of the existing literature concentrating on these countries has focused on the potential for Nordic (and now Baltic) cooperation. In this chapter I argue that attention should now be more firmly placed on the (often) neglected elements of intra-state regionalism taking place in these countries as well.

Regionalism and Local Government in Sweden

The intention of this chapter is therefore to focus on intra-state regionalism as applied to the Scandinavian and more specifically Swedish case. To aid in this short survey, the difficult distinction between 'regionalism' and 'regionalisation' should also be mentioned. Authors such as John Loughlin have argued that 'regionalism' refers to a bottom-up process – an ideology and/or political movement advocating greater control by regions over the political, economic, and social affairs of their regions, usually in the form of political and administrative institutions with legislative powers. In contrast, and not necessarily connected to 'regionalism', 'regionalisation' is an approach defined by central government or supranational bodies (e.g. the European Union) from the point of view of the polity as a whole. It refers to the process by which central authorities define and redefine their relationship with the component regions within nation state territories and may or may not include the devolving of powers by the centre to the regions (Loughlin, 1998).

In addition, Sweden has been described elsewhere as something of an enigma

(Miles, 1997a). In geographical terms it is the fifth largest country in western Europe, covering 449,964 square kilometres. However, with a population of 8.9 million, it also enjoys an exceedingly low population density, only 21 inhabitants per square kilometre. It is dominated demographically, economically and politically by its three largest (and southern) cities of Stockholm, Göteborg and Malmö, which act as the focal points for economic and political activity. In a country of such geographical size, it is by no means surprising that there are clear regional and cultural differences between the essentially rural and sparsely populated North and the more cosmopolitan and densely populated South. The southern county of Stockholm, for example, is home to 1.7 million Swedes, compared to the northernmost (and largest) county of Norrbotten with a population of only 267,000. However, in terms of 'regionalism', the impact is mixed. Given increasing volatility and scepticism amongst voters towards the mainstream national parties, there has been an increasing number of local-based parties (Petersson, 1994, p. 130). The proportion of the total number of votes for these parties at local elections has risen from 0.5 to almost 3 per cent in the last 20 years. Minor parties are currently represented on a third of Sweden's municipal councils. At various times minor parties such as the Stockholm Party (1979) and the Scania Party in Malmö (1985) have held the balance of power on 'hung' councils. At the same time, there have been few distinct (local) political or ideological movements (apart from the constituency associations of the national parties) in Sweden and certainly no separatist movements since the break-up of the 1905 union with Norway. Federalism as a concept has also found little favour in Sweden.

This is not to say that regional interests are not also important to national political parties in Sweden. Some of the newly prominent parties, such as the Greens and Christian Democrats (since 1988) have substantial presence and electoral bases at the local level, and the agrarian Centre Party has usually been more popular in the rural regions. Swedish governments have also been sensitive to the concerns of the Northern peoples, using regional aid as a means of maintaining population levels in the rural Northern communities. However, in general, 'regionalism' has been limited in the country and the loyalties of the Swedish voters to the existing state, and especially crown, remain strong. Nevertheless, this has not prevented the Swedish state from having a long tradition of local autonomy, with formal separation of ecclesiastic and secular affairs taking place as part of the local government ordinances of 1862. Formal concepts of 'regionalisation' are still relevant to the Swedish case and in the post-war period, local government has assumed a growing importance. Although the 'great reform of local government' in the 1950s halved the previous number of 2,500 municipal authorities or communes (to 1,037), the expansion of the welfare sector in Sweden in the 1960s and 1970s ensured that municipalities and county councils assumed ever growing significance. By 1974, boundary reforms were completed and smaller municipalities were merged into large 'super-municipalities' reducing the total number to 278. Since then, several of these municipalities have been partitioned into two or more on the basis of demographic change, bringing the total to the present 288.

The growth of the public sector continued more slowly in the 1980s and, although facing greater financial difficulties in the 1990s, the role of local government has not diminished. In fact, as Sweden still maintains a large public sector, with public sector expenditures equivalent to around 68 per cent of GDP in 1995, local government continues to remain important as a focus for public services provision.

Broadly, there are two layers of local government (excluding parish councils). These are the regional county councils (*landsting* – sometimes referred to as 'secondary municipalities') and the smaller ('primary') municipalities (*kommun*) – a status maintained in the 1991 Local Government Act (which acts as a *de facto* 'constitution' for Swedish local government at municipal and county levels). These 288 municipalities and 23 county councils, covering 24 counties, account for 25 per cent of GNP and 30 per cent of all Swedish jobs (1995 figures) and are broadly responsible for around 70 per cent of public sector activities (excluding transfers). The municipalities provide services, such as planning, the maintenance of the physical environment, education, public health, emergency services, civil defence, transportation and communications, technical services such as water, sewage and energy, and recreational and cultural programmes. The main task of the county councils is to oversee the health care systems in their territories, which makes them slightly unusual in the sense that, unlike county councils in other countries, they do not have a broad range of responsibilities. Nevertheless, given that these councils are accountable for a large proportion of the public sector through their supervision of health services, they should not be dismissed out of hand. Indeed, their significant role is indicated by the reduced level of political tension between the main political parties in the running of the councils. The services that they provide are deemed to be so important as to be largely outside day-to-day political dogfighting, and decisions of the county councils are often consensus-driven.

Taking account of the wealth of services traditionally provided by an essentially social democratic-orientated state to its citizens, it could be argued that local government provides the principal point of contact between government and the average Swede. However, the regional level is not usually considered to be the most important stratum of government (Larsson, 1995, p. 73). With the possible exception of the activities of the county councils, the regional level is usually the weak link between the national and municipal levels. In most cases, the main responsibility for implementing measures falls either at the central or local (rather than at the regional) level. Most activities take place in the municipalities. They therefore enjoy a dual role as sovereign decision-makers on certain matters and the executors of state decisions on others.

Local government has also enjoyed relatively high levels of importance precisely because of the unique way Swedish public administration operates. While the country is widely classified as a unitary state, in which the powers of the government and Riksdag are, more or less, identified in the 1974 Instrument of Government, Sweden has for three centuries had a bipartite structure for public administration. Matters which in other countries would usually belong to a ministry are divided between Swedish ministries and public authorities, such as central administrative agencies like the National Labour Market Board, the National Agency for Education and the National Board for Health and Welfare. Not only this, but the principle of ministerial rule is not applied, ensuring that administrative authorities enjoy a large degree of independence, whilst being accountable to the Swedish Cabinet as a whole. The organization of central administration is therefore complex, leaving clearer lines of accountability at the local level and strong attachment amongst Swedish citizens to the notion that public services are provided by local government.

In addition, Sweden also has a strange mixture of functional and prefectoral systems of local government (Larsson, 1995, p. 73), with aspects of both systems used simultaneously. Occasionally, some of the central public agencies have regional organizations (i.e. according to the functional principle). On the other hand, central government also maintains a presence through the office of County Governor (*landshöding*) and the county administrative boards (*länsstyrelse*). In spite of the fact that these are regional bodies, they are directly responsible to government (resembling the prefectoral system), even if by no means all central authorities act through county administrative boards. The office of the County Governor, first instituted in the 17th century, acts as the representative of the national administration, and oversees the implementation of government decisions at the local level, such as the collection of taxes. Similarly, the county administrative boards implement objectives that have been determined at the national level throughout the respective county. However, the county administrative boards are not just the mouthpiece of the state at the local level; they have also evolved a valuable role in promoting the interest of the county at the national level (Petersson, 1994, p. 101), tasks which suggest that the County Governor becomes a representative of his/her region rather than an executor of government policies.

The Challenges Confronting Regionalism in Sweden in the 1990s

As in other European states, Sweden's regions and local government are responding to changes emanating both from within Swedish society and from outside in the 1990s. They are dealing with competing pressures from which, to a large extent, they remained immune in previous decades:

Financial austerity and deregulation

The first of these is without doubt the financial problems of the Swedish public sector in the 1990s, which inevitably impinge upon local government in its role as the main provider of public services to citizens. The expansion of local government operations paralleled general economic growth during the 1960s and 1970s, enabling the local tax base to be extended in line with growing prosperity levels. However, this trend came to a halt in the 1980s and, since 1985, state grants have declined in value.

Since the early 1990s, local government has been confronted by demands from the centre to reduce public expenditure and levels of provision at a time when high levels of unemployment in Sweden have placed even greater demands upon local services. The Government and Parliament have been in general unwilling to countenance increases in municipal resources as a means of returning the national state budget into the black. The national budget was running a deficit at nearly 15 per cent of GNP in 1993 (and only after drastic cuts will return to surplus by 1998).

Since the severe recession of 1990-93 and the relatively fragile economic recovery following it, local tax revenues have grown slowly or not at all, ensuring that local government revenues are falling in fixed price terms (Häggroth *et al.*, 1996, p. 107). In addition, central government has financed local government principally through block grants rather than selective state grants since 1993, enabling the centre to save money. These problems have been further compounded by the general demographic trends in Sweden, with a growing segment of the population over 65 and requiring extensive and expensive support from local health care and social services. A 1998 report to the Swedish Finance Department estimated, for instance, that the costs of caring for the elderly were likely to rise steadily until the year 2020 and to explode by 2030, with the total costs for elderly care 60 per cent higher by 2030 (Regeringskansliet, 1998, 15, 130).

The solution to the dilemma of greater pressure on public services, especially as more Swedes exhaust their unemployment benefits, and the municipalities become responsible for longer-term social assistance, has been through a process of deregulation of services and the introduction of 'purchaser/provider internal markets' within public services. This of course is by no means new as the social democratic governments began a limited injection of deregulation into public services in the 1980s. Yet, the process has been substantially accelerated, with some services, such as the provision of homes for the elderly, being partially supplied by private firms in the late 1990s (Svenska Kommunförbundet, 1995b).

Internationalisation: European Union membership and the Baltic connection

This general process of deregulation has been accompanied by the growing internationalisation of Sweden and its economy. In particular, Sweden's membership of the European Union (since 1 January 1995) creates new opportunities, but also imposes new limits on the municipalities and county councils.

On the one hand, European Economic Area (initially) and then European Union membership has first, reinforced the ethos of deregulation and greater competitive pressure in public service provision and secondly, presented greater opportunities for developing regional cooperation within Sweden. Taking the former, European Union rules on opening public procurement contracts for local services, have been (for the most part) adopted into Swedish law (since joining the European Economic Area in January 1994). The Swedish European Union Consequences Commission set up to estimate the potential impact of European Union membership on the country argued in 1993 that the municipalities and county councils could reduce their expenditure on tenured contracts for public services by 10,000 million SEK (European Union Consequences Commission, 1994: 30) as a result of adopting European Union rules. However, large areas of public services will remain untouched by Swedish European Union membership. The Union presently has few powers to regulate education, social services or health and medical care provision, which are large areas of local governmental remits.

Perhaps what is more important is that the Union has promoted keenly the concept of a 'Europe of the Regions' and encouraged regional authorities to work together. The southern regions of Sweden (led by Skåne) for example, have realized the potential benefits of attracting European Union funding to improve links with neighbouring Denmark, including the financing of the Öresund bridge-tunnel project, which provides a new transport link between southern Sweden and the major Danish conurbation. Moreover, one county (Jämtland) and parts of a further six counties, covering about 50 per cent of Swedish territory (241,640 square kilometres,) and 17 per cent of the population (449,000 people), now qualify for European Union regional aid under Objective 6 (based on low density of population – under 8 persons per square kilometre) of the Structural Funds. Indeed, it is both ironic and by no means accidental that the regions qualifying under Objective 6 are those with the populations that voted most heavily against European Union membership in the 1994 referendum. It remains to be seen whether awarding them European Union funds will be a way of changing their minds in the future!

Not only are there potential financial rewards in Swedish regions collaborating together when bidding for European Union Structural Funds, but the municipalities have also realized the political advantages in lobbying European Union institutions and utilizing their representation on the European Union's Committee of the Regions. The Swedish Association of Local Authorities has operated a special international unit since the early 1990s which, amongst other things, coordinates the Association's lobbying activities in Brussels. The Association has, for example, highlighted the potential influence given to municipalities through their representation on working groups of the European Commission and the Committee of the Regions. Without doubt, Swedish municipalities and county councils will become even more internationally orientated in the future, both in terms of bilateral relationships with foreign counterparts and exchange of services and expertise.

However, European Union membership does raise some important questions for Swedish regionalism. To a degree, the Swedish public administration tradition can be seen as unique, and given the lack of a convincing definition of a region at the European Union level, Swedish government may be forced to reinvent statistically congruent geographical units so as to enable comparisons with other European regions to be made. The fear within parts of Sweden is that European Union requirements will have a centralizing, rather than decentralizing effect, as new 'informal' regions complicate the existence of the country's strongly autonomous local government. The local authorities also remain suspicious of the European Union's attempts at harmonizing taxation and insistence on stringent fiscal measures as part of the European Union's single currency programme, since local government is responsible for providing large areas of Sweden's expansive welfare services. In particular, the reform of the Swedish VAT system in order to adapt to European Union rules and the fact that local taxation makes up 54 per cent of local government revenue, means that local authorities are sensitive to European Union policies that require changes in the level and basis of Swedish public expenditure. To offset these concerns, great stress has been placed by local government on two aspects (Svenska Kommunförbundet, 1995a) – first, the European Union's continued emphasis on the 'subsidiarity' principle (in which policies are made at the level that is the most efficient, but also more importantly at the level closest to the citizen); and second, that levels of public access should be improved through an adoption of rules of transparency and openness in European Union decision-making (included in the 1997 Amsterdam Treaty).

Of course, regional collaboration is not something new in the Nordic countries. Regional projects have been financed previously under the auspices of the Nordic Council since the 1950s (Thomas, 1996). Yet the areas adjacent to the Nordic countries, the Baltic Sea and the Arctic including the Barents Sea, are also opening up and exerting a major influence on the neighbouring regions in Sweden. New bodies such as the Council of the Baltic Sea States and the Barents Council have promoted regional initiatives in which the respective Swedish regions are encouraged to participate (Archer, 1998). It would seem at least, that there is now a greater potential for Baltic or European Union sponsored initiatives involving Swedish regions, rather than purely Nordic cooperation. The combined influences of European Union membership and closer Baltic rather than purely Nordic initiatives, have widened considerably the opportunities for regional cooperation encompassing Swedish municipalities and county councils.

Changing public values and attitudes

Social scientists have also argued that changing attitudes and even values, although hard to quantify, have also affected the role of local government. Clearly, one of the most discernible changes in Swedish society over the last 20 years has been the greater emphasis placed on 'individualism'. Hence, whilst various public opinion surveys, such as those by the SOM Institute at Göteborg University, have highlighted declining public confidence in Swedish governmental and parliamentary procedures (Miles, 1997b), the same cannot necessarily be said for attachment to regional and local government which has remained fairly consistent. Although still essentially a social democratic society, the greater emphasis on deregulation by government and individualism by elements of Swedish society has strengthened citizens' loyalties to their local communities. Moreover, since the 1991 Local Government Act gave local government greater powers over, for example, the formulation of their committee structures, this tendency has continued. Thus, 85 per cent of Swedes consistently indicate that they receive good services from their municipality (Häggroth *et al.*, 1996, p. 110), even if they express considerably lower levels of confidence in actual local politicians. Support for services remaining at the local level is therefore strong.

However, at the same time, greater market orientation of local government operations has altered their role as 'service providers'. Increased pressure on local government to seek greater efficiencies in service provision (without raising taxes) and more competition to improve choice and local consumer power, ensure that Swedish local government, like most local government throughout Europe, will gradually be one of many, rather than the sole service provider to the local communities. As Häggroth *et al.* have commented, the debate on 'the limits of local self-government' continues in Sweden, even if at a slightly slower pace than in other countries such as the United Kingdom.

Conclusion

The 1990s will, one suspects, go down in the history books as being a particularly significant decade in the development of the Swedish state and regions. As Marie Pernebring commented in 1995, 'There is a wind of change blowing in the local government sector in Sweden' (Svenska Kommunförbundet, 1995b, p. 5). The dual pressures of internationalisation from above and individualism from below have rocked traditional concepts of social democratic Sweden. While not in any way meriting the removal of the 'social democratic label' from this country, successive governments have nonetheless, in the 1990s, abandoned the last vestiges of a 'Swedish model', and moved headlong along the road toward greater liberalisation and a deregulated state. In a geographically diverse country such as Sweden these changes at the central government level inevitably have implications for the nation's regions, especially when local government has traditionally been the supplier of extensive public services during the peak days of the welfare state. It would seem that the Swedish welfare state is, to a larger extent than ever before, run and administered by the regions and municipalities.

However, although there are greater pressures on the regions and local government in particular, these competing trends of internationalisation and individualism do represent potential opportunities. As in other countries, Swedish local government has not been slow in recognizing the financial and political possibilities offered by Sweden's accession to the European Union, even if traditional forms of Nordic cooperation will most likely take place increasingly within wider Baltic and European Union initiatives. The process of deregulation and market orientation, originally instigated by central government and reinforced by the European Union, will continue unabated in Sweden. The challenge for individual Swedish regions will be to ensure that they are the ones within the country to benefit most from the new opportunities created.

References:

Andrén, N. (1967), 'Nordic Integration' *Cooperation and Conflict,* 2 (1), 1-25.

Archer, C. (1998), 'The Baltic-Nordic Region', in Park, W. & G. Wyn Rees (eds), *Rethinking Security in Post-Cold War Europe,* Harlow, Longman, pp. 117-34.

Childs, M. W. (1980), *The Middle Way on Trial,* New Haven, Yale University Press.

Häggroth, S., K. Kronvall, C. Riberdahl, & K. Rudebeck, (1996), *Swedish Local Government: Traditions and Reforms,* Stockholm, Svenska Institutet.

Larsson, T. (1995), *Governing Sweden,* Stockholm, Statskontoret.

Loughlin, J. (1998), 'The Regional Question In Europe: An Overview', Paper presented to the 'Federalism and European Union' Conference, Centre for EU Studies, University of Hull (March 1998).

Miles, L. (1997a), *Sweden and European Integration,* Aldershot, Ashgate.

— (1997b), 'Sweden: A Relevant or Redundant Parliament, *Parliamentary Affairs,* 50 (3), 423-37.

Petersson, O. (1994), *Swedish Government and Politics,* Stockholm, Publica.

Svenska Kommunförbundet (1995a), *Developments in the European Union and their Effects on Sweden's Local Authorities,* Stockholm, Svenska Kommunförbundet.

Svenska Kommunförbundet (1995b), *Waves of Renewal: New Trends in Public Management and Administration in Swedish Local Authorities in the Nineties,* Stockholm, Svenska Kommunförbundet.

Thomas, A. H. (1996), 'The Concept of the Nordic Region and the Parameters of Nordic Cooperation', in Miles, L. (ed.) *The European Union and the Nordic Countries,* London, Routledge, pp. 15-31.

Wendt, F. (1979), *Cooperation in the Nordic Countries,* Copenhagen, The Nordic Council.

Regionalism in Italy

Anna Bull

Introduction

The Italian regional question is threefold. It is important to distinguish between regions as administrative entities, regional policy, including regional development, and regionalism/federalism as a political movement and current of thought. In Italy all three have played a part in shaping national as well as local politics since Unification, although at different times and with varying fortunes. Indeed, it is only in recent years that these different aspects of regionalism have all become deeply enmeshed and achieved prominence at one and the same time, the catalyst for this being the formation of new regionally-based political 'Leagues'.

The main issues arising from the Italian regional question are the following – first, in terms of the regions themselves, the main issue at the time of Unification was whether to introduce a system of administrative centralisation or decentralisation. Despite the pressure put on the Government by some advocates of a federal State, the question of whether to grant the regions autonomy rather than some measure of administrative devolution of power was never seriously considered. By contrast, the issue has now shifted dramatically and revolves around the alternatives of introducing a federal system of government or granting the regions considerable further administrative and fiscal powers, eg. the right to impose their own taxes.

Second, in terms of regionalism/federalism as a current of opinion capable of influencing party politics, the issue at the time of the Risorgimento was how best to reconcile and integrate so many different peoples and cultures. Today, after more than 130 years of political unity, a much higher degree of homogeneity has been achieved, but from the point of view of socio-economic development (and in very recent times also in terms of political behaviour) the country appears to be divided into three inter-regional 'blocks' – the North, the Centre and the South. Federalism has greatly revived, thanks primarily to the rise and success of the Northern League party, and it has recently influenced all main political parties, although there is much uncertainty, as well as disagreement, over the meaning of the term.

Third, In terms of regional policy and regional development, the issue at the turn of the century was how to achieve a redistribution of resources in favour of the poorer and less developed regions of Italy, i.e. the South. The North was criticised for draining resources from the South through an unfair taxation system and spending a higher proportion of public funds in the more affluent North. After the Second World War the issue became how best to promote economic development in the South through the channelling of specially ear-marked State funds. Today it is the very idea of State-funded regional development which is in question.

To enable the reader to follow the historical development of these three aspects of Italian regionalism from Unification to the 1980s, I will first consider them separately. When discussing present-day Italy, however, I will consider regionalism as a single, though multi-faceted, issue precisely because, as I mentioned earlier, all the above aspects seem to have become inextricably linked.

Historical Survey

The regional system and the Unification period

At the time of Unification, the Italian Government was faced with the dilemma of administrative centralisation or decentralisation. In those days 'Piedmontisation', i.e., the hurried extension of Piedmontese legislation to the newly annexed Italian regions, was resented in the North as well as in the South, although perhaps not to the same degree. The Piedmontese and the Lombards were different peoples with very different political-historical experiences, and so were the Tuscans, Emilians, Sicilians, Neapolitans etc. Only 2.5 per cent of the population knew Italian at the time of Unification, a figure that includes the Tuscans, upon whose dialect the national language was based (De Mauro, 1963, p. 43).

The diversity of Italy's component regions as well as growing resentment in the country against Piedmontisation convinced Cavour as well as many other Italian political leaders that some measure of devolution ought to be granted. Cavour himself was a believer in decentralisation although, in the words of Mack Smith, 'he hardly had time to make up his mind' (Mack Smith, 1968, p. 341). In this, as in other instances, the Liberal Governments of Italy genuinely professed certain ideals but in practice had to introduce something very different when faced with the reality of Italian society and politics.

A scheme of regional devolution, the Farini-Minghetti bill, was prepared in 1861 and approved unanimously by the Cabinet but later withdrawn when it became clear that centrifugal forces, particularly in Southern Italy, could jeopardise the newly unified kingdom. Cavour himself changed his mind after Unification, shortly before his death in June 1861 'despite the fact that he continued to deplore centralisation as illiberal, expensive and inefficient, he had been compelled to modify his views when he saw the danger that Italy might fall apart if a uniform administrative system was not quickly imposed on the whole kingdom' (Mack Smith, 1985, p. 263).

The most pressing agenda for the Italian ruling class at the time was how to 'harmonise' regional differences. In this context both the supporters of a centralised State and those of a federal State (see 'Regionalism as a political movement and current of thought' below) had a common aim, although they differed in what they saw as the means to achieve this aim, alternatively 'from above', i.e., through the imposition of a uniform and centralised State apparatus, or 'from below', i.e., through a slow process of amalgamation and progressive elimination of local/regional differences. That the solution adopted was centralisation from above should be seen as a measure of the

weakness of the Italian agrarian and industrial bourgeoisie which had been the driving force behind Unification and its inability to impose cultural and political hegemony over society as a whole.

Thus in place of the Farini-Minghetti bill the Government passed a Law in 1865 (Law N. 2248) which introduced a rigid prefectorial system along Napoleonic lines. The prefect became the representative of executive power at local and provincial level with wide-ranging authority over numerous spheres of influence, including education, law and order, administration and justice.

Establishing the regions

It was not until the end of the Second World War that administrative decentralisation was once again seriously considered. One of the reasons for this was a general agreement that Fascism's rise to power had been made easier by the centralist character of the Italian Liberal State. A more balanced division of power would prevent the recurrence of an authoritarian solution. Another powerful motive force was the climate of reforms prevalent after the war, which led to a widespread consensus that the reconstruction of the Italian political system ought to take place along new democratic lines and not be remodelled around pre-fascist Liberal institutions.

Despite these common aspirations of the anti-fascist parties, there was no clear convergence on the question of regional autonomy. The Socialist and Communist Parties, in particular, were suspicious of any form of federalism in case it promoted reactionary political tendencies at the periphery – the left-wing parties in this respect showed only limited appreciation of the innovative potential of a regional political system (Ragionieri, 1976, p. 2481).

The end result was that the Italian Constitution, elaborated in 1947 and formally introduced on 1 January 1948, established the regions as administrative entities with limited legislative powers in a number of fields, including police, health services, town planning, tourism, local transport and communications, public works, agriculture and forestry. The regions were denied 'primary' legislative responsibility, i.e. the authority to legislate independently of the State and were attributed only 'concurrent' and 'subsidiary' legislative responsibility, in other words the authority to formulate legislative initiatives complementary to or within the framework of national legislation. The 1948 Constitution provided for the establishment of 20 regions, of which five were to enjoy 'special autonomy' (or 'Statute', equivalent to a region's constitution) and the remaining 15 'ordinary autonomy' (or Statute).

In this, as in other fields, the Italian Constitution was not applied for several years. The Statutes of four of the five special regions (Sicily, Sardinia, Valle d'Aosta and Trentino Alto Adige) were approved in February 1948; the fifth special region, Friuli-Venezia Giulia, was established in 1963. There were specific political reasons why these regions received favourable treatment, namely the fact that they included considerable ethnic minorities (the three Northern ones) or had shown separatist tendencies (Sicily and Sardinia).

The 'ordinary' regions had to wait until the 1970s, despite the left-wing parties' change of heart vis-à-vis decentralisation after they were ousted from government in

1947 and after their defeat in the general elections of 1948. The Christian Democratic Party [Democrazia Cristiana, or DC] was in power and had no intention of strengthening the Communist Party [Partito Comunista, or PCI] in the latter's regional strongholds in Central Italy. Nor were there strong demands for decentralisation coming from Italian society. According to Nanetti, 'it was a society with a small number of élites whose economic and political interests were served well by centralised institutional decision-making [...] These demands [for regional Government] had to wait for the political events and economic and social changes of the 1960s' (Nanetti, 1988, p. 80).

Thanks to these changes, above all the development of an increasingly pluralistic society with the emergence of new social and interest groups accompanied by the political success of the Left parties, pressure mounted in the country for breaking the Christian Democrats' monopoly of political power and for a degree of power-sharing. The creation of the regions in 1972 was granted by the then Prime Minister Andreotti in response to this new social and political pressure. Great hopes were expressed at the time regarding the regenerating effect administrative decentralisation would have upon the Italian system of Government, which by then had shown itself unable to implement radical reforms in line with the country's rapid industrialisation. Yet cautionary notes were sounded in various quarters. Earle summed up the various reactions to the regional experiment at the beginning of the 1970s as follows, 'At best it can inject new vigour and more direct democracy into the machinery of government, acting as a vehicle for progress and enrichment of life at all levels. At worst, it can insert a parasitic layer of maladministration between the central government and the ninety-four provinces, adding to the opportunity for *clientelismo*, intrigue and corruption' (Earle, 1974, p. 90). As we shall see, both forecasts turned out to be correct.

The regions since 1970

A series of laws were passed in the 1970s to set up the regional system. The 1972 decrees followed the 1970 regional elections granting the regions limited powers. Law 382 of 1975 gave the regions wide-ranging powers within the scope of the Constitution (no primary legislative authority therefore). The 616 decrees in 1977 'institutionalised the regions as real centres of policy-making' (Nanetti, 1988, p. 81). The decrees gave the regions control over 25 per cent of the entire national budget (Putnam, Leonardi & Nanetti, 1985, p. 80). It should be noted, however, that according to Article 119 of the Constitution the regions enjoy 'financial autonomy in the forms and within the limits prescribed by the Republic's laws which coordinate it with the finance of the State, Provinces and Communes'. The prevalent interpretation of Article 119 has always been that the financial autonomy of the 15 ordinary regions is very limited, as they cannot impose new taxes or regulate the imposition or distribution of taxes which are already in place. Thus the 'financial autonomy' attributed to the regions by Article 119 consists mainly in the autonomy to administer directly their income as fixed by the State. Greater financial autonomy is attributed to the five regions with special Statute, although only in exceptional cases do they have the power to impose their own taxes.

As for regional government, Article 121 of the Constitution states that 'the organs of the Regions are the Council, the Junta and the President', whose functions correspond roughly to those of Parliament, the Government and the President of the Republic. The council is made up of between 30 and 80 councillors and has legislative, regulating and administrative responsibilities, as well as functions of political control over the junta and the president and also over the region's policy-making. The junta is the executive body and is elected by council, whereas the president, also elected by council, is both the region's representative and the president of the junta. Councils approve the regions' Statutes, which require an absolute majority and have to be approved by Parliament by law. The Statutes regulate the internal organisation of the region, including the composition of the junta and the system for electing the junta and its president. They also regulate relations between the different regional Government bodies and their functions. Regional Statutes, however, cannot regulate or modify the type or number of regional Government bodies, the functions of such bodies, the electoral system for councillors or the widening of a region's territory. Statutes also contain regulations concerning popular legislative initiatives and regional referenda.

Below the regions, administrative decentralisation rests with the provinces and the communes. The provinces, after the creation of the regions, became rather hybrid institutions, since many of their powers were lost. Many experts advocated the abolition of the provinces but a recent Law, passed in 1991, retained them as administrative entities (see below). The communes are the only territorial bodies to pre-exist Italy's Unification and indeed the Constitution simply granted them official recognition, as opposed to creating them *ex novo*.

Since the establishment of the regions, the two main areas of contention and ambiguity have been precisely the relation between, and respective spheres of influence of, the regions and the State on the one hand and the regions and the other tiers of local Government (provinces and communes) on the other (Cammelli, 1990). As far as the regions and the State are concerned, the main limitation to the autonomy of the former is the Government's power to contest the constitutionality of a regional law. The Government can invite the regional council to reconsider any law this has passed; such a law can be re-approved by council, provided it is approved by an absolute rather than a simple majority of councillors. After that, central Government has 15 days in which to appeal to the Constitutional Court. Conversely, the regions themselves can resort to the Court against the State if in their opinion it violates their functions.

There have been numerous cases brought to the attention of the Constitutional Court since the regions started to operate. The Court's initial tendency was to rule in favour of the State, thus in the 1970s the prevalent attitude on the part of the regions was to find a compromise with the State and avoid recourse to the court for fear of this body's anti-regional orientation. (Rodotà, 1986, p. 91). In the 1980s the attitude of the Court towards the regions became much more positive. In the same decade, regional autonomy vis-à-vis the State increased considerably so that this aspect can no longer be treated as a central issue; rather, it has been replaced by a struggle between regional and local government (Putnam, 1993, p. 46).

Have the regions found popular favour? Have they commanded attention and established roots? These questions need to be considered alongside the question of the performance of the regions, which also achieved new prominence since the overcoming of the main problem posed by State-regions relations. Both the above questions, i.e. popular affection/disaffection towards the regions and regional performance, are closely linked, since new institutions need to have a positive impact upon the society in which they operate in order to establish roots and command popular support.

According to Hine, the performance of all regions has been disappointing in so far as they have become an integral part of the Italian political system and are in themselves entrenched in 'partyocracy'. 'The parties act as channels through which Regions and regional party leaders can bring pressures on the centre' (Hine, 1993, p. 271). Despite their limited financial autonomy, for example, the regions operated large deficits in the 1980s in the knowledge that the centre, and the national taxpayers, would bail them out, which regularly happened. From this point of view the pessimistic forecast that the regions would add another layer to the clientelistic Italian body politic can be deemed to have been correct. In the early 1990s, a corruption scandal known as 'Tangentopoli', from the word 'tangente' meaning a cut paid by private and public companies to political parties in exchange for public contracts and favourable treatment, involved many local and regional executives throughout Italy. The scandal was uncovered in Milan but it soon spread to other parts of Italy, involving all the main political parties, above all the Christian Democrats and the Socialists [Partito Socialista, or PSI]. Whereas scandals of this type had previously often been associated with party politics and State intervention in the South, Tangentopoli seemed to have unified all Italy on the basis of maladministration.

Beyond and above the clientelistic aspect typical until recently of Italian politics at all levels, it is possible to distinguish between two different groups of regions in terms both of performance and popular esteem. Various studies have brought to light the consistently better performance of the Central and Northern regions vis-à-vis the Southern ones (Leonardi, Nanetti & Putnam, 1985; Putnam, 1993). The perceptions of the performance of local and regional government on the part of Italians vary sharply, accurately reflecting this geographical division. Northern and Central Italians are generally satisfied with their local and regional governments whereas they express dissatisfaction with central Government; by contrast, Southern Italians are dissatisfied with all tiers of government, which they judge to be inefficient, ineffective and corrupt (Putnam, 1993, pp. 54-6). The reasons for this are varied and have been traced back to different levels of economic development, uneven distribution of resources, historical traditions and, more recently, different degrees of 'civicness', measured in terms of active participation in democratic and political associations, trust and solidarity, community values and political equality (Putnam, 1993, pp. 86-120).

As these issues concern the social and political cultures of different regions of Italy, rather than the regional system *per se* and bring back to prominence the Italian 'Southern Question', they will be discussed in some detail below, together with the new regionalist political movements and approaches to regional policy and regional development.

Regionalism as a political movement and current of thought

The federalist movement at the time of Unification

Strictly speaking, one cannot speak of a federal political movement at the time of Unification. The two great political leaders of the Italian Risorgimento, Cavour and Mazzini, were both anti-federalist, although they were in favour of some measure of devolution of power. There were, however, various federalist thinkers who were in varying degrees influential among the political and cultural elite, despite lacking a popular following. Among these one should mention Carlo Cattaneo (1801-1869) and Giuseppe Ferrari (1811-1876), both from Lombardy. The former was a radical Liberal who believed that individual freedom came before nationalism and the respect of local/regional diversities before the need for a strong, centralised State. Before 1848 Cattaneo was ready to accept autonomous rule for Lombardy within a federal Austrian Empire (Mack Smith, 1968, p. 92) and even after that date he was more concerned with the abolition of trade barriers between the various Italian regional States than with national unity. Cattaneo was not a political leader and his influence on Italian society and politics remained limited, although in Lombardy itself he was a popular figure. Likewise, Ferrari was an intellectual and philosopher who remained at the margins of active politics. He, too, advocated a federal Italy, on the grounds that 'our history rejects the possibility or desirability of our becoming a unitary nation; on the other hand a federal system will enable us to reach the very highest goals [...] We may regard federalism as the purest form of constitutional government, founding liberty on a written pact, on a multiplicity of assemblies, on the inviolability of all internal frontiers, and the solemnity of its central Parliament' (Speech to Parliament, 8 October 1860).

Vincenzo Gioberti (1801-1852), a Piedmontese, advocated a federal union sponsored by the Pope, a project which collapsed when Pope Pius IX turned against Liberalism and the cause for National Unification after the 1848-49 revolutions. Gioberti had a clear vision of the fact that an Italian people did not exist and needed to be created taking into account existing divisions in 'government, laws, institutions, popular folklore, customs, sentiment and habits'. To this end he judged the creation of a single unitary State to be either madness or, if brought about by force, an immoral crime.

As Donzelli (1992, p. 6) pointed out, the federalist current of thought that can be traced back to the Italian Risorgimento tradition can be defined as 'instrumental in ascent', that is to say, a federalism which identifies strong ethnic, linguistic, social and cultural differences within a territory seen as capable of achieving national unity. The federal system is seen in this context as a transitory system, an instrument to bring about the 'harmonisation' and real unification of a State's regional components.

The new federalism

In the late 1980s and early 1990s federalism became the battlecry of the Italian Northern League Party [*Lega Nord*]. It differed fundamentally from that of the Risorgimento and was defined (Donzelli, 1992, p. 6) as 'instrumental in descent', despite the League's protestations that they were Cattaneo's natural heirs. By this expression Donzelli referred to the League's secessionist aspirations and its openly-held conviction that

Northern and Southern Italy represented two distinct and non-converging societies which ought to be free to go it alone. The League's position presupposes the total rejection of the Italian Fathers' aspirations to achieve complete unity through the creation of a common people.

Paradoxically, the revival of federalist/ethnic sentiments took place at a time when Italy has reached a high degree of cultural homogeneity, not least from a linguistic point of view. Apart from minority ethnic groups, linguistic unification is now an accomplished reality. Census results indicate that Italian is now prevailing, although the dialects have not disappeared. Most people can speak both Italian and a dialect, and the percentage of people who speak only or mainly Italian is constantly growing (Lepschy & Lepschy, 1991; Lepschy, Lepschy & Voghera, 1993).

The Northern League strongly defended the use of the dialect and provocatively asked for the 'Lombard' dialect to be used as official language. Yet it was not linguistic 'nostalgia' that provided the stimulus for the new federalism. Rather, it was the deterioration of political institutions, the growth of organised crime, and the systematic use of corrupt practices in business transactions involving party and State officials unearthed by the recent 'Tangentopoli' scandal.

In the eyes of the League the 'Southernisation' of the Italian State as evidenced by the emergence of a widespread corrupt and clientelistic system of Government was a clear sign that the process of national Unification, far from promoting the homogenisation of the country around 'Northern' laws, practices, institutions and economy, had succeeded only in imposing 'Southern' deviant practices upon the whole of the country.

In addition, socio-economic development remains uneven. The North/South divide is, as we shall see, still highly relevant, but there are also differences between the other regions. Back in the 1970s Bagnasco (1977) identified 'three Italies' in terms of social and economic structures – the industrial, urban-centred and large-firms-dominated North-West; the newly developed, still semi-rural, small-firms-dominated North-Eastern and Central regions; and the underdeveloped South. Since then much important research has been produced attempting to explain the reasons behind, as well as the characteristics of, each area (or model) of development.

The federalism of the League, as put forward in the party's 1992 electoral programme, appeared to have been rather crudely inspired by the findings of Bagnasco and other sociologists. The League advocated the creation of a Federal State made up of three macro-regions or Republics (North, Centre and South), each considered as homogeneous from a socio-economic point of view. The Federal State would be responsible only for foreign affairs, defence, justice, general finance and higher education. The emphasis was on the creation of a Northern Republic made up of Lombardy, Piedmont, Venetia, Liguria Emilia-Romagna and Tuscany, represented in the party literature as the most socially advanced part of Italy, governed and 'oppressed' by the Southern-dominated State bureaucracy and party system.

The starting point for setting up the three macroregions, or at any rate the Northern one which was the one that really mattered for the League, was, according to the party, a formally legal and constitutional procedure. Article 132 of the Constitution states that it is legitimate to provide for the fusion of existing Regions by Constitutional Law, as long as this is requested by a number of Municipal Councils representing at least a third of the interested population, and provided that the proposal is approved in a referendum by a majority of the same population.

The Northern League was considered in the late 1980s a fringe protest movement, with an extreme populistic political programme which would fail to make inroads into the more educated and middle-class electorate. However, this prediction turned out to be wrong. The party obtained good electoral results in the Northern provinces, particularly in Lombardy, at the 1990 administrative elections, did surprisingly well in the 1992 general political elections and won across the Northern regions in the 1993 local and provincial elections, gaining Milan as well as scores of minor wealthy industrial towns.

The collapse of the Socialist and Christian Democratic parties following the *Tangentopoli* scandal appeared to have created a new division in the country in terms of political behaviour, with the possible consequence of a more drastic political split. In 1994, however, a new party, *Forza Italia*, was set up, led by the media tycoon Silvio Berlusconi, which put national unity and even nationalism back on the political agenda. Another party, *Alleanza Nazionale*, made up largely of ex-fascists, also stood for national unity. These two parties combined with the Northern League to form an uneasy coalition, the *Polo delle Libertà*, which formed a government after the March 1994 political elections. A few months later, the Northern League withdrew its support from this government. In 1996, new political elections took place, resulting this time in the formation of a centre-left coalition, the *Ulivo*, which proceeded to form the next Government. Since then, the Northern League has moved decisively towards outright secessionism, while the other parties have ostensibly embraced federalism in one form or another, not least in order to steal the thunder from the Northern League itself. One of the possible stumbling blocks on the road to a federal state remains the question of the North/South divide. I will now discuss this in greater depth.

Regional policy and regional development

The Italian Regional Question centres around the underdevelopment of the South, and both regional policy and regional development schemes have been dominated by the need to solve this basic North/South divide. The first serious attempt to promote economic and social development in the region occurred in the first decade of the 20th century, when successive Italian Governments, headed by Giuseppe Zanardelli first and Giovanni Giolitti later, prepared a series of laws designed for a few southern regions. At that time Italy had begun to industrialise (the country's first industrial take-off is deemed to have occurred between 1896 and 1913). Provision was made for measures of agricultural improvement, public works, and health schemes. As for industrial development, the city of Naples was designated as an industrial growth area and a new steel plant was set up in one of its suburbs. The policy largely failed, partly due to the absence of an indigenous entrepreneurial culture but above all, according to Clark, to a lack of resources as well as corruption: 'here are the beginnings of constant themes in twentieth-century Italian politics: the distribution in the South of subsidies and patronage by central State development agencies, and the use of such agencies to win political support' (Clark, 1983, p. 133).

In the 1950s the Italian State made a much more decisive attempt to develop the southern economy. In August 1950 the Law establishing the *Cassa per Opere Straordinarie di Pubblico Interesse nell'Italia Meridionale*, better known as the *Cassa per il Mezzogiorno*, was passed. The function of the *Cassa* was clear from its title, namely to undertake 'extraordinary interventions' over and above what normal Government Ministries could achieve. In the intentions of the legislators the *Cassa* was to be a public body with its own legal status, largely independent of the civil service. Yet it was made responsible to the Minister for the Mezzogiorno and subject to Government control. The *Cassa*'s efforts to develop the South can be divided into three main phases – 1950-7 when the emphasis was on modernising agriculture and building infrastructure related primarily to agriculture; 1957-71 when industry was singled out (Law 634 of 1957) as the sector that needed to receive greatest attention; 1971-84, i.e., the post-'heavy industry' and recession years.

The first phase was marked, as well as by the creation of the *Cassa*, by the land reform laws of 1950 – *Legge Sila* in May, *Legge Stralcio* in October and *Legge Siciliana* in December. These laws were aimed at reducing the power and the size of the estates of the large absentee landowners (*latifondisti*). Part of their land was expropriated and subdivided into small plots of land which were assigned to landless labourers or petty landowners. The farming plots thus created had the short-term effect of reducing the chronic unemployment and underemployment of the area and easing social tensions, as well as stealing the thunder of the Communist Party, which had backed the southern peasants' agitations of the 1940s and received growing political support from them in return. The policy of favouring the peasant farm model, however, was not economically viable and when Italy joined the European Community the backward state of southern agriculture became even more evident. Emigration to the North at the time of the country's economic miracle also drained human resources from the southern countryside and resulted in the abandonment of many of the newly-created family farms.

The shift in regional policy from promoting agriculture to promoting industrial development was thus inevitable. What kind of industries should be promoted became a key issue. An important objective of regional development was to encourage the formation of small and medium-sized firms and the emergence of indigenous entrepreneurship. In reality, however, the Italian State had to rely on the public sector, mainly the two giant State holding companies IRI and ENI, to make investments in the South, with the consequence that many industrial plants set up in the southern regions were in the capital-intensive, 'heavy' industrial sector, whose success turned out to be deeply conditioned by external events. With the energy crisis of the early 1970s and steel overproduction throughout Europe the fate of these plants became highly uncertain. Furthermore, there were few 'trickle-down' effects upon the local/regional economy and these large plants' failure to stimulate the growth of small and medium-sized firms earned them their famous nickname 'cathedrals in the desert'.

Lack of coordination and planning, excessive bureaucracy, and sheer corruption were blamed in the 1970s for the poor results achieved by the *Cassa*. Excessive centralisation was another factor judged as having a negative effect on southern development. After the creation of the regions in the 1970s, pressure mounted for delegating some or all of the powers of the Central Agency to the regions; indeed in the late 1970s there were increasing demands for the suppression of the *Cassa*. The *Cassa* was finally abolished in 1984, and in 1986 Law 64 gave extraordinary powers to the regions in what constituted another major policy shift. It was felt that if regional governments were left to formulate and implement their own plans, intervention would be far more effective as the regions were more familiar with local conditions, needs and resources.

One of the main problems identified with the functioning of the *Cassa* was the gap between legislation and its transmission and translation into action. This has been largely explained on the basis that the *Cassa* was used to fulfil a political function – local Christian Democratic leaders systematically plundered resources to give their party a strong power base (Chubb, 1990). In this context it mattered little whether the plans were actually implemented, so long as money continued to sustain the party's own clienteles. In the 1980s the Socialist Party exercised a similar role.

Law 64 of 1986 was subjected to intense scrutiny and it incurred severe criticism. The regional governments proved largely incapable of drawing up, let alone implementing, development projects themselves, and the usual problems of clientelism and corruption were still apparent, this time on a regional level. Regarding the maladministration of the southern regions, regional policy and regional development have not yet found their best administrative/political framework.

Following the collapse of Italy's First Republic, the very concept of regional development was called into question, despite the ever-present gap between North and South. This was partly due to the fact that the failure to develop the South in the past decades led many Northerners to doubt the wisdom of subsidising the South (Becchi, 1992). Less 'noble' motives included fiscal pressure (harsher in the North), the huge Government deficit and the economic recession which hit Northern industry in the late 1980s.

A study promoted by the Regional Council of the Veneto Region showed that between 1985 and 1990 four Northern regions, Lombardy, Piedmont, Veneto and Emilia-Romagna paid 45 per cent of national taxes, 62 per cent of VAT, and 63.5 per cent of local taxes. They were given by the State 33.9 per cent of the funds redistributed to local and regional governments. Thus for every 100 Lire paid to the State the Lombards received back for their own use 24.5, the Piedmontese 30, the Venetians 35, and the Emilians 37. By contrast, Molise could spend 80, Campania 64, Puglia 58, Calabria 84 and Basilicata 85 (*La Repubblica*, 1 April 1992, p. 14). The Southern regions depended on Government financing and 'are terrified that the Government may ultimately expect them to depend on their own resources and taxes instead of giving them lavish subsidies' (Haycraft, 1985, p. 19). As Hine explained, in Italy's regional system the coalition between the Ministry of Finance and the Southern regions had numerical superiority over the Northern regions, which were often politically divided among themselves. The new federalism advocated by the Northern League in the early

1990s must thus be viewed in the context of this increasing frustration and sense of impotence on the part of the richest half of the country.

Italian Regionalism Today

The Evolution of Decentralisation

In June 1990 a new Law on local Government (Law N. 142) was passed, with two primary objectives. The first objective was to reduce clientelism and encourage local Governments' accountability to their electors, mainly through greater participation by ordinary citizens (*via* referenda, petitions or proposals – Article 6), access to information (Article 7) and the creation of a '*difensore civico*' or guarantor of impartiality and good administration (Article 8). The second objective was to regulate relations between the three tiers of Government, central, regional and local. Greater importance was attributed to the regions in this tripartite relationship compared to that allowed for by Articles 117 and 118 of the Constitution. Whereas these articles stated that 'Provinces and Communes are autonomous bodies within the context of the principles established by the laws of the Republic which determine their functions', Article 3 of Law 142 elevated the regions to a position of superiority vis-à-vis the provinces and communes in terms of socio-economic and territorial planning. The regions were assigned the important functions of determining general policy objectives, although local and provincial Governments were to contribute to their formulation. Law 142 was in this respect ambiguous since it did not clarify in what ways and on the basis of which criteria communes and provinces contributed to general policy planning.

Despite the wider functions attributed to the regions by the new Law, they were far from being satisfied. The rise of the Northern League Party and the extreme form of federalism it advocated provided the regions with both new ammunition and a new sense of urgency. On 19 December 1990 the regional council of one of the regions with special Statute, Friuli-Venezia Giulia, approved, with the support of the Christian Democrats, Socialists, Communists and Greens, an agenda for a federalist reform of the Italian State. A similar request was put forward by Trentino-Alto Adige on 19 February 1991. On 8 March it was the turn of another 'special' region, Valle D'Aosta. On 26 February 1991 one of the 'ordinary' regions, the wealthy Northern Emilia-Romagna, requested a new type of regionalism. The main target of all the above regions was Article 117 of the Constitution, which limits the functions of the regions.

The mood of the Italian people, meanwhile, appeared to have turned decisively in favour of the regions, if only in protest against the maladministration of central Government. On 18 April 1993 Italians were asked to vote in eight referenda, two of which were sponsored by the regions. One of these concerned the abolition of the Ministry of Agriculture, the other the abolition of the Ministry of Tourism; in both cases the functions of the Ministries were to pass to the regions. Both referenda were approved by a majority of Italians.

A third referendum was to determine whether Italians wished to extend to all towns the majority system which assigns two-thirds of the seats to the winning list and to elect the mayor directly in municipal elections. This referendum was also approved, together with a fourth which established that 238 Senators out of 315 were to be elected with a majority system and only 77 with the system of proportional representation established in Italy after the fall of Fascism. These last two referenda represented the clearest indication yet of the will of the Italians to change their electoral and political system at the roots and to create their Second Republic.

Against this background, a further move towards greater administrative decentralisation took place in 1997, when two Laws, Law 59 of 15 March and Law 127, of 15 May, the so-called 'Bassanini' Laws, were introduced by the centre-left government. These Laws represented, on the one hand, a continuation of the trend in the direction of conferring wider functions and responsibilities to the regions in the context of a unitary state, already established in 1990. On the other hand, they also introduced new, potentially radical, principles. In terms of the trend towards further decentralisation, Article 2 of Law 59 for the first time listed the functions which were reserved for the central state, leaving all other areas under the jurisdiction of the regions and local councils. The list, however, was criticised for being overlong – no fewer than 20 areas of policy were reserved for national government, including controversial ones such as telecommunications and the protection of the artistic heritage (Rugge, 1997, p.721). In addition, the 1997 Laws retained the same ambiguity as Law 142 of 1990 concerning the relationship between the regions and the lower tiers of local government and as such they were subjected to the same type of criticism. In particular, it seemed that the legislators were primarily concerned with ensuring a high degree of consultation and cooperation between the various layers of government at the expense of effective decision-making. According to Rugge, however, the 1997 Laws also introduced two radical new concepts. The first was that of 'subsidiarity', the second, potentially more important in the Italian case, was that of 'differentiation', i.e., the idea that the degree of local and regional autonomy granted to regions and local councils ought to vary in relation to their concrete administrative skills and performance (Rugge, pp 723-24).

Almost as soon as the 1997 Laws were introduced, the debate moved on, focusing increasingly around federalism rather than administrative decentralisation. With the exception of the Northern League, the positioning of the other political parties started to converge in the works of the Bicamerale, as we shall now see.

Regionalism as a political force

Today the Northern League has shifted position. It now advocates secessionism. As Gilbert put it 'By the election campaign of 1996 [...] the last nod in the direction of national unity had disappeared, and the word 'Padania' had entered into the Italian language. The Lega's choice of foreign model, it had become apparent, was Czechoslovakia' (1998, p. 52). This left the other parties free to occupy the 'space' left vacant by the League, thus (re)appropriating federalism. As Vassallo (1997, p. 694) wrote, 'At the time of the 1992 elections, only the representatives of the League were"federalist" [...] By the 1996 elections all the main political groups had become convinced supporters of federalism, even though the term was generally linked to adjectives such as "solidarist" or "unitary"'. At this stage, 'federalism' started to mean all things to all parties. At the end of January 1997, the Bicamerale, or Bicameral Committee for Institutional Reform, was established by act of parliament (Gilbert, 1998, p. 53). A first proposal, delivered on 4 November 1997, was largely in conformity with the 'Bassanini' Laws, and for this reason it attracted similar criticisms (Vassallo, 1997). While it claimed to represent a decisive move in the direction of a federal state, with the central State left with a coordinating role and the regions entitled to financial autonomy and full powers of intervention in all sectors within their territories, it was judged to be simply administrative decentralisation by another name. More recently, in April 1998, the Chamber of Deputies approved Articles 57, 58, and 59 of the new Constitution, reducing from 20 to 10 the areas of policy attributed to national government and re-affirming the principle of 'differentiation'. Reactions to these articles were somewhat more positive than they had been at the end of 1997, although many commentators still maintained that the State was preserving a key role as mediator between the Regions and the other tiers of local government. Even more recently, the Bicamerale appeared on the verge of collapsing, with the serious possibility that institutional reform might be postponed indefinitely.

It therefore remains extremely difficult to predict how the country's political institutions will be reformed, including the regional system. Leaving aside the secessionist demands of the League, a federalist solution along German lines seems to date the most probable outcome of Italy's revived regionalism. However, the temptation on the part of Italy's new political class to fudge the reform of the organisation of the state and to establish a new equilibrium based on consensus, rather than competition, between the various political forces and tiers of government, should not be underestimated.

What future for regional development?

Whether a more extreme or a more moderate form of federalism will prevail (or simply a more substantial devolution of power to the regions), the Italian regional divide is clearly there to stay. The question is, will a federal State or a more radical form of regionalism promote or hinder the cause of national harmonisation? Would it simply represent the triumph of the new 'selfishness' of the richest regions or could it also be considered as a step forward for the least developed ones? Is there scope for renewed, state-led schemes of regional development?

More than 40 years of regional policy and regional development have left the South at best modernised but certainly not developed. In the early 1990s there seemed to be a growing consensus that there must be an end to indiscriminate subsidies and politically-motivated, centrally-controlled transfers of money from the northern to the southern regions. The League's solution would be a neo-liberal one, where incentives to development should consist of lower wages (justified on the basis of an alleged lower cost of living in the South), tax incentives for businesses on the basis of strictly documented profits on their part, and an end to all State subsidies. The traditional parties were more oriented towards the preservation of some forms of direct subsidies, but with more rigorous methods of control.

There seems to be no agreement as to the best way forward. The poor performance on the part of the Southern regions so far leaves serious doubts about their ability to make use of a higher level of political autonomy – but a lower level of financial means – in order to promote development. On the other hand, so many paths have already been followed that there is a growing current of opinion that the South will either find its salvation from within or not find it at all. In this context the regional/federal solution is seen by many as a new beginning for what will no doubt be a painfully slow process of institution-building and socio-economic development. Conversely, others stress the need for central government to regain the initiative and re-launch a comprehensive plan for economic development, rejecting any neo-liberal solution.

The centre-left government has so far made timid steps in both directions. It has gone some way in the direction of increasing the degree of flexibility of the labour market, even going as far as accepting *de facto* the reintroduction of some wage differentials between North and South as an incentive for firms to operate in and move to the South. At the same time, the government seems not to be averse to the idea of recreating a special Agency for regional development with responsibility for the South. An articulated and multi-dimensional approach to this question, devoid of dogmatism, may yet prove the best way forward, although there are risks of contradictory policies and a lack of firm direction.

Italian Regionalism within the Context of Europe

Italy was a founder member of the European Community and the Italian population has consistently showed firm support for European integration (Hine, 1993, p. 286). European unity is not a political issue in the country; if anything, the recently formed Northern League is even more pro-European than the traditional parties. This is due to the fact that they believe that a Europe of Regions will eventually supplant the existing Europe of Nation States and thus allow northern Italy to rejoin its transalpine neighbours.

As Umberto Bossi stated in his autobiography 'What is the meaning of having frontiers between Piedmont and Savoie, or South Tyrol and Austria? Their ethnicity is substantially identical, from a naturalistic point of view. From a socio-cultural point of view [...] nothing unites Trentino or Lombardy with Calabria or Campania. Therefore I say: why not replace the fixed frontiers and the centralism typical of unitary States with a more articulated system, characterised by a plurality of institutional centres each with specific and limited responsibilities? Why not eliminate, in other words, the rigid frontiers between very similar realities, as for example Lombardy and Bavaria, while introducing separate decision-making centres, each with real autonomy, in different realities which were arbitrarily unified, such as the North and South of Italy?' (Bossi, 1992, pp. 161-2).

The advocation of a return of northern Italy to its 'natural' Mitteleuropean cradle could not have been more openly and clearly stated. As with the League's other political proposals, such ideals reflect the complete failure, in their eyes, of the Italian nation-state. European integration is viewed in this context as little more than an opportunity to achieve the party's secessionist goal.

The possibility that Italy might not be among those countries which would be able to join the Single European Currency in the first round, coupled with the risk that Italy's exclusion would simply fuel support for secessionism in the North, greatly influenced the policy agenda of the centre-left government. In 1996 and 1997, effective measures were taken to reduce the country's excessive deficit, keep a tight rein on expenditure, and control inflation. By 1998, it became clear that Italy 'had made it', and that she would join EMU when it was first introduced. This should help reduce internal regional and political tensions.

The pro-Europeanism of most Italians, however, had little to do with perceived inter-regional and trans-national social and economic homogeneity and a lot to do with a distrust of central Government, i.e., the same distrust which turned Italians into supporters of a radical new form of regionalism. As *The Economist* pointed out in 1993, 'Italy was an enthusiastic signatory of the Maastricht Treaty in December 1991, not just because it has always been Europhilic but because the Maastricht requirements of economic and monetary convergence would impose the discipline that Italy's governments normally fail to find on their own [...] What will happen now, if the Maastricht process becomes more fuzzy? The answer is that Italians will have to rely on their own politicians, not anyone else's' (*The Economist*, 26 June 1993, p. 21).

Now that Italy is part of the EMU project, Italians can arguably continue to rely upon Europe to provide an impulse for institutional and political modernisation. Conversely, a radical and effective reform of the country's political system and the establishment of a strong and authoritative executive may yet re-establish the credibility of the Italian nation-state in the eyes of its citizens. Were this to happen, the merits of a united Europe, whether federal or confederal, would be assessed and judged in their own rights and not as providing a substitute for weak national Government.

References

Bagnasco, A. (1977), *Tre Italie. La problematica territoriale dello sviluppo italiano*, Bologna, Il Mulino.

Becchi, A. (1992), 'Incentivi statali, pesante eredità', in *A sud di qualunque nord*, Il Manifesto del mese, n. 8.

Bossi, U. (with D. Vimercati) (1992), *Vento dal Nord. La mia Lega la mia vita*, Milan, Sperling & Kupfer.

Cammelli, M. 'Regioni e rappresentanza degli interessi: il caso italiano', *Stato e mercato* , N. 2 (August 1990), 151-200.

Chubb, J. (1990), *Patronage, power and poverty in Southern Italy*, Cambridge, Cambridge University Press.

Clark, M. (1983), *Modern Italy 1871-1982*, London & New York, Longman.

De Mauro, T. (1963), *Storia linguistica dell'Italia unita*, Bari, Laterza.

Delpino, L. & F. Del Giudice (1992), *Diritto amministrativo*, Naples, Simone.

Donzelli, C. (1992), 'Le oscillazioni del federalismo', in *A sud di qualunque nord*, Il Manifesto del mese, n.8.

Earle, J. (1974), *Italy in the 1970s*, Newton Abbot & Vancouver, David & Charles.

Falzone, V., F. Palermo & F. Cosentino (eds) (1976), *La Costituzione della Repubblica Italiana. Illustrata con i lavori preparatori*, Milan, Mondadori.

Gilbert, M. (1998), 'Transforming Italy's Institutions? The Bicameral Committee on Institutional Reform', *Modern Italy*, 3 (1), 49-66.

Haycraft, J. (1985), *Italian Labyrinth*, Harmondworth, Penguin.

Hine, D. (1993), *Governing Italy. The politics of Bargained Pluralism*, Oxford, Clarendon Press.

Italia, V. & M. Bassani (eds) (1990), *Le autonomie locali. Legge 8 giugno 1990, n. 142)*, Milan, Giuffrè.

La Repubblica, 1 April 1992.

Lega Lombarda-Lega Nord, *Programma elettorale* , February 1992.

Lepschy, A. L. & G. Lepschy (1991), *The Italian Language Today*, London, Routledge.

Lepschy, A. L., G. Lepschy & M. Voghera (1993), 'Linguistic Variety in Italy', paper given at the 1992 Annual Conference of the Association for the Study of Modern Italy [ASMI]. A summary of this paper appeared in the ASMI Newsletter, N. 23, Spring 1993.

Mack Smith, D. (1985), *Cavour*, London, Methuen.

— (ed.) (1968), *The Making of Italy*, New York, Harper & Row.

Nanetti, R. (1988), *Growth and Territorial Policies. The Italian Model of Social Capitalism*, London & New York, Pinter.

Putnam, R. D. (with R. Leonardi & R. Nanetti) (1993), *Making Democracy Work. Civic Traditions in Modern Italy*, Princeton, New Jersey, Princeton University Press.

Putnam, R. D., R. Leonardi & R. Nanetti (1985), *La pianta e le radici*, Bologna, Il Mulino.

Ragionieri, E. (1976), *La storia politica e sociale*, Storia d'Italia, Vol 4:3, Turin, Einaudi.

Rodotà, C. (1986), *La Corte Costituzionale*, Rome, Editori Riuniti.

Rugge, F. (1997), 'Le leggi "Bassanini": continuità e innovazioni del riformismo amministrativo', *il Mulino*, 4 (July-August), 717-26.

Rupeni, A. (ed.) (1980), *Comuni e Province negli anni 80*, Rome, Cinque Lune.

The Economist, 26 June 1993.

Vassallo, S. (1997), 'Il federalismo sedicente', *il Mulino*, 4 (July-August), 694-708.

Regionalism in Greece

Dimitrios Christopoulos

The Greek Regions

Greece is a country of approximately 10.5 million inhabitants who live in a terrain dominated by mountain ranges traversing the country from North to South and an archipelago of thousands of islands, 300 of which are inhabited. The morphology of Greece is obviously partly responsible for development differentials, as the various mainland regions and islands have different levels of accessibility, transport infrastructure and communications links. Only about 17 per cent of the country is suitable for arable crops, with 40 per cent classified as pasture and 20 per cent as woodland which gives an indication of the wide variety of regional needs and capabilities.

Physical barriers to interaction are not alone in causing the great disparities in rates of development and growth among the Greek regions. One of the most inhibiting factors for regional development has been the extreme centralisation of the Greek state. In the last few decades the state, motivated mainly by an economic logic, has attempted to counteract the decline in the productive capacity of regions by attempting mainly to deconcentrate government departments, while since the mid-1980s an un-concerted attempt has been made at decentralising decision-making to the regions. Since the mid-1990s, however, with the election of regional prefects and the reorganisation of local government, there is evidence of the beginning of a regionalisation process, partly affected by the need of the state to modernise and partly by the effects of the Europeanisation of the policy process.

To assess the relevance of regionalism in Greece we have first to understand the parameters behind the creation of the contemporary socio-political landscape. In the mainly agrarian Greek society urbanisation was a result of lack of employment opportunities in rural areas but also of the civil war (1946-9) that followed the Second World War. Fear of the 'communist threat', among the civil war victors, created the backdrop for a pathological rejection of any points of view that did not conform to the mainline ideology. This is hardly the spirit that would promote the advancement of local democracy and instead created a background of massive migration to urban centres where 'left-inclined' individuals escaped to avoid harassment. The net effect of the economic and political pressures on rural communities was that Greece experienced unprecedented urbanisation, with half of the country's inhabitants living in one of the two main urban agglomerations and all major cities under the continuous pressure of ill-planned expansion.

Urbanisation has also been aided by industrialisation. During the 1950s and 1960s rapid industrialisation allowed Greece to aspire to joining the developed world. Industrial production accounts for 25.3 per cent of Gross Domestic Product (including mining and construction) but manufacturing is only responsible for 14.3 per cent of GDP. In 1993 services accounted for 67.2 per cent of GDP while agriculture for 9.6 per cent. The large and rather inefficient agricultural sector (although declining to 7.3 per cent of GDP by 1995) is an impediment to growth, particularly if compared to a European Union average of 1.8 per cent. The biggest weakness of the Greek economy however, is perceived to be the large size of the public sector, estimated at 65 per cent of assets (EIU, 1998). Although a programme of gradual privatisation is in hand, political pressures and entrenched labour interests are unlikely to allow the programme to bring about any significant results over the next decade.

A positive note on the prospects of the Greek economy can be sounded from the comparatively high GDP growth over the last few years (3.5 per cent in 1997) and the historically low inflation rates attained recently (5.5 per cent in 1997). GDP per head reached $11,426 in 1997, hovering around 63 per cent of the European Union average. Transfers from the European Union represented 30 per cent of invisible receipts in 1997. In 1994 Greece contributed 1.5 per cent to the Community finances and received 8.0 per cent of Community payments, a net transfer of 3812.6 million ECU. However, in spite of this huge transfer of funds, the deficiency of the Greek economy in attracting foreign direct investment (FDI) and the persistence of a structurally uncompetitive industrial base is behind estimates of negligible effects from the Single European Market (ESRI, 1997) for the Greek economy. A similar picture emerges in the evaluation of the effect of the Community Support Frameworks (Delors I and II operational between 1989-1999) where GDP is not expected to grow by more than 1 per cent by 2010 beyond what would have been the case if the programmes were not in place. This suggests that the significant transfers from the European Union may not necessarily have a long term effect and would not necessarily narrow intra-European Union disparities. This pessimistic picture could change for the better if some of the recent economic measures (such as joining the ERM in March 1998), were to increase confidence in the Greek economy and therefore have an effect on FDI.

In recognising the economic inefficiencies entailed in allowing internal regional disparities in Greece to aggravate, the state has created a number of incentives, including tax breaks and grants for companies with operations in the periphery, while the Hellenic Industrial Development Bank (ETVA) finances 20 industrial estates in outlying prefectures. Regional disparities are very pronounced, with the region of Sterrea Ellada being close to 75 per cent of the European Union GDP average, while Ipeiros is close to 42 per cent.[1] Some of these regional disparities will be aggravated by the new focus of the main infrastructural investments over the next decade. Of major importance is the organisation in Athens of the 2004 Olympic games. This has resulted in a growth in government investment for the capital with the creation of two additional metro lines, a new airport and the upgrading of motor and rail links. A number of regions that are within reach of the rail and motorway construction will benefit directly, but an even greater number are potentially deprived of infrastructural

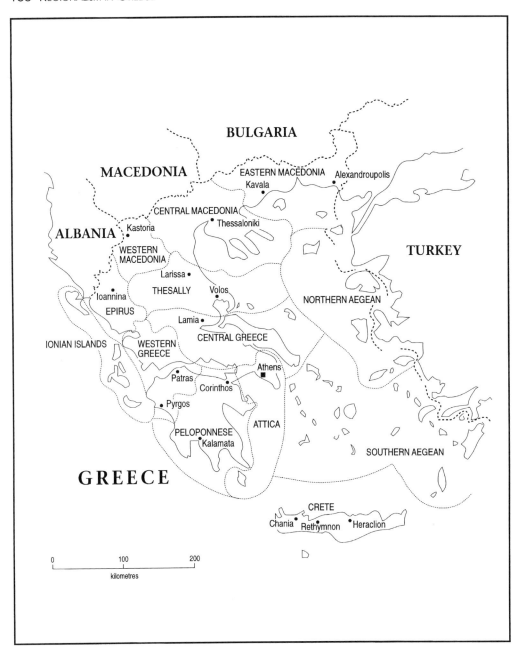

investment. The Greek administration still appears to lack strategic vision in regional planning. Verney noted that in the 1980s 'no regional development plan existed, while national five year plans...[lacked] a guiding and unifying philosophy' (1994, p.170). Very few things appear to have changed in the 1990s. The same author also suggests that the causes of regional neglect might be more sinister as 'assigning a lower priority to regional development is likely to reduce the pressure for political reform' (Verney, 1994, p.177).

The Modern Greek State

The accession of Greece to the EEC in 1981 is a pivotal point in the gradual modernisation of the Greek state. The main challenge from this point onwards was the adaptation to the *acquis communautaire*. Initially this was an effect of becoming eligible for structural fund support, that included transfers from the Common Agricultural Policy (CAP), the Social Fund and the Regional Fund. The participation in regional and structural policies was soon coupled with challenges of institutional transformation that included the ratification of the Single European Act (1986), the instigation and attempt at absorption of funds from the Integrated Mediterranean Programmes (1985) and the effort to conform with the European directives of the common market project (1992). All these events produced great challenges for the state administrative structure and the political establishment.

During this period the administration faced its greatest challenge since the creation of the modern Greek state in 1831. The predominant logic in the formative years of the state was one of centralisation, since in the 19th century Greece was rapidly expanding its territory and a number of challenges to the state were underpinned by instances of localism. Centralisation was therefore identified with modernisation. The effectively napoleonic administrative institutions guaranteed a comparatively efficient control of regional political elites and the containment of corruption and of arbitrary administrative actions at the local level. This centralisation however quickly led to clientelism and administrative centralism. In periods of authoritarian rule these conditions were aggravated.[2]

Political traditions were first challenged seriously by calls for administrative devolution in the early 1980s, as centralism was perceived to be causally related to the vertical clientelistic networks proliferating between the administrative and political centre in Athens and the regional urban and rural areas. Lyrintzis (1984) coins the term 'bureaucratic clientelism' to describe a fusion of the structures of the party in power and the state apparatus in post-1974 Greece. The objective of such a fusion is the clientelistic distribution of the 'spoils of government' to the party faithful. Diamandouros suggests that underlying the weakness of civil institutions and civil society has been a profound individualism as well as the imposition of capitalist institutions to a pre-capitalist society that created a 'deeply alienative political culture' (1983, p. 48). Political parties and the administrative elites are therefore the primary agents of the corruption of democratic institutions and highly unlikely to promote any measures that will deprive them of any of the political spoils. Lyrintzis defines the limits of bureaucratic clientelism to be in 'the presence of well-known politicians heading strong local or regional factions among their ranks' (1984, p.104). Localities can therefore be seen to exhibit some resistance to the influence exerted by the national party machinery. Although it could be argued that local networks of clientelism and corruption are simply contesting national ones for political space, their existence could also be seen as the context within which the propagation of local value systems has survived the contestation from the central state. Furthermore, it could be argued that exploitation from national clientelistic networks could provide the background within which local and regional societies will become increasingly alienated from the nation state and the backdrop from which demands for regional devolution may emerge in the future. This will be the case particularly if regional political elites become more 'emancipated', with their actions determined by local rather than national considerations.

Accession to European Community institutions and accruing benefits from programmes such as the IMG's and the Common Agricultural Policy brought with them demands for probity by the European Commission and other Community partners. As the fund transfers increased commensurate with further European integration, their significance in economic terms (5 per cent of GDP by 1994) forced the Greek political class to take heed of the European Commissions' anticipation that funds should be administered at the regional level. Or, to be more precise, the European Council regulation assumes that the implementation of regional programmes like the IMPs will be 'at the relevant geographical level' (EEC/2088/5/July 1985). Further pressure has been applied in the 1990s by an effort to establish a so called 'partnership' principle between the Commission and European regions.

Attempts at Administrative Decentralisation

It has been argued that the Socialists in the mid-1980s were attempting to fulfil an electoral pledge to decentralise the state in order to increase administrative efficiency and reduce patronage. It was perceived that such changes would provide an economic lifeline to the Greek periphery and increase citizen participation and levels of local democracy. Failure of the socialists to change the administrative structures is an indication of the strength of the established patronage networks and of political opportunism in exploiting those networks for political gain, as I have argued elsewhere (Christopoulos, 1998).

Change was brought about by a multitude of factors. The European Commission and the European Economic Community/European Union partners pressed Greece to adopt administrative modernisation. Politicians in urban centres resented the dominance of the Athenian administrative and political elites. Opportunities in the Balkans and Eastern Europe (particularly since 1991), by increasing contacts and creating the opportunities for externally generated growth, have enhanced the assertiveness of elite actors in northern cities such as Thessaloniki, Kavala, Ioannina, Kastoria and Alexandroupolis. Economic and political elites in other significant economic urban or urban-rural agglomerations are also gradually acquiring a degree of political assertiveness. Most notably in the cities of central Greece such as Larissa,

Lamia and Volos; the Peloponnesian cities of Corinthos, Patras, Pyrgos and Kalamata; and the Cretan cities of Heraclion, Rethymnon and Chania. Finally, and most significantly, the need to speed-up economic growth while at the same time reducing the size of the state (particularly if Greece was to follow its partners to a Monetary Union) meant that successive administrations were desperate to find efficiency gains even at the risk of a major overhaul of administrative structures.

As a result of this pent-up pressure for modernisation the Socialists, under Andreas Papandreou, introduced a Bill on 'local government, regional development and democratic planning' (Law 1622/86) that stipulated a degree of administrative autonomy for the Greek prefectures by introducing elected councils. Although this Bill passed through Parliament and was enacted into Law, elections to the Councils were not called for another eight years! Under what is perceived to have been extensive political manoeuvring the 'Law in limbo' was revised in 1989 (1832/89), while eventually in 1994 a number of Bills proclaimed the 'establishment of prefectural self-government' (Laws 2218/94 and 2240/94) and provided for the election of prefects (*nomarches*) and prefecture councils. Until then prefecture chairs and councils were political appointees of the government in power. Elections to the 52 prefectures were first held on 16 October 1994. Significantly and perhaps dubiously the Law affirms prefectures 'do not have a superintendent role over local authorities' and that there is 'no hierarchical relationship between the two tiers' of local government and prefecture councils (Law 2218/94, section1, p.1269). The state in other words retains the role of ultimate arbiter of both tiers of government for itself. At the same time the current administration wishes to be perceived as seriously dealing with the decentralisation agenda. A signal to that effect was the recent renaming of one of the main government departments to Ministry of Interior, Public Administration and Decentralisation.

Almost all commentators therefore agree that the national political elite is cautiously proceeding to a rationalisation of its administrative structures, not as a result of pressure from below but rather in order to modernise and conform to European directives.[3] The latest attempt at this rationalisation process has been the successful introduction of the so called Kapodistrias Law (2539/97) of 1997, named after the first Greek governor, which reduced the 6,000 odd local authorities, most of which were small villages, to just under 900 municipalities and local communities. This was done by forging neighbouring communities to consolidate their administrative structures. Interestingly, most resistance to this Law came as a result of the disruption of local clientelistic networks, while protests had only a very small effect on eventual mergers. The process of administrative transformation is also expected shortly to lead to an elected regional tier for the 13 regional cabinets and chairs (*periferiarhes*) who at the moment are political appointees. We can only speculate as to whether the new institutions will lead to a genuine 'emancipation' of regional units or will instead lead to fragmentation of local demands by tiering administrative structures between local, prefecture, regional and national.

Effects of Europeanisation

As the regionalisation of Greece is a *sui generis* process there are a number of issues on which it is difficult yet to make a pronouncement. It is too early, for instance, to determine all the effects current administrative changes will have on local identity and political culture. Some evidence suggests that the strength of regional identity has implications for the political and economic interaction in certain Greek regions (Konstadakopulos & Christopoulos, 1997). It has also been argued that localism is a facet of Greek political culture inextricably linked with the birth of the modern Greek state (Diamandouros, 1983; Clogg, 1992). It is therefore likely that administrative devolution will eventually lead to a reassertion of local identity. However, this does not necessarily imply a 'contest' for the nation state. There appears to be no significant conflict in the hierarchical structuring of identities. Regional identities can be perceived to belong to a concentric set of mental maps where the regional is plainly subjugated to the national sense of identity (for a Cretan example see Herzfeld, 1985).[4] It could, on the other hand, be argued that Greece will follow the example of Italy and Spain in their experience of regionalism. As Ritaine argues, civil society may be essential in this process. Maybe the pluralist and neo-corporatist conceptions do not in themselves provide adequate explanations of regional dynamics. In that respect the hypothesis of the political exchange paradigm might be relevant. 'Confronted with the weakness of state regulation these [regional] societies have produced forms of political regulation' (Ritaine, 1998, p.87). It is not improbable that we will witness a delayed effect from the same parameters that affected the Spanish and Italian regions on some of the Greek regions as well. For those theorists for whom southern Europe represents a 'distinct and identifiable space' this may indeed be a real possibility. Bottom-up regionalism in that event may emerge due to the weakness of the state to regulate efficiently and effectively. However, there is no evidence to suggest that regional identities could be contesting national ones. Therefore, if a regionalist rhetoric was to emerge it is very unlikely that it would be aimed at that part of the nation state that underpins the foundations of national identity; it would most likely be aimed instead at the inefficiencies and corruption of the central administration.

For students of European regionalism Greece represents a classic case of a unitary, highly centralised state which, in an attempt to increase internal administrative efficiency and in response to the Europeanisation of its policy domain, has been led to regionalise. The process has been a slow one and, as Christakis argues, 'Greece has adapted faster politically than administratively' (1998, p.98) to European Community institutions. Indeed the national political elite seems embedded in the process of political and economic integration as there appear to be significant direct and indirect gains derived from membership. It is easy to deduce therefore that calls for increasing local democratic accountability have at best been relevant to central government's decentralisation rhetoric and not at any point been the driving force behind regionalisation.

Regionalism vs. Regionalisation

Latent tensions from perceived economic exploitation and political domination of the periphery by the centre cannot at this stage be seen as politically significant. The future of the relationship between Athens and the Greek regions will depend on a great variety of factors including:

- the fate of the vertical clientelistic networks between centre and periphery,
- the speed of modernisation of the national administrative apparatus,
- the speed with which regional administrations and local government will adapt to the transformations at hand,
- the Europeanisation of Greek politics and administrative practices,
- the transformation of institutional relationships between all tiers of government and the European Union,
- the extent of transformation of European Community institutions and the pressures that will emanate from these for the Greek administration,
- the effect economic development will have on political demands,
- regional political entrepreneurship,
- the assertiveness of regional identity and political culture and
- the ability of the state to 'preside' and control a multi-tier structure of local-prefecture-regional-national-European governance and policy making.

Whichever way one examines this gradual transformation of the Greek state, evidence presented here suggests that the process of decentralisation has given way to a phase of regionalisation that is bound to affect irrevocably the relationship of the periphery with the national centre.

Notes

1. During the last few years the economic landscape has also been affected by a great influx of economic migrants who provide cheap manual labour. These are mainly workers from Albania who, according to some accounts, now exceed 600,000. Analysts predict their presence will have some positive impact on the productivity of certain agricultural regions and a negative impact in the investment climate of some northern regions.
2. For instance during the Papadopoulos/Ioannides dictatorship (1967-74) local government officials were appointed by central government.
3. Among the critics of the Greek administration are Andrikopoulou (1992), who talks of a failure effectively to utilise regional elites by giving them power for local development planning, and Ioakimidis (1996), who argues that the state continues to be centrist. Theodorou (1995) gives a comprehensive account of the history of local government institutions in modern Greece, while Kostopoulos (1996) discusses the possibility that in Greece we witness a case of 'decentralised centralisation'.
4. This may not be the case for the Muslim minority in Thrace, who however represent less than 1 per cent of the population of Greece.

References:

Christakis, M. (1998), 'Greece: Competing Regional Priorities' in Hanf, K. & B. Soetendorp (eds), *Adapting to European Integration; Small States and the European Union*, London, Longman, pp. 84-99.

Christopoulos, D. (1998), 'Clientelistic Networks and Local Corruption: Evidence from Western Crete', *South European Politics and Society*, 3 (1), 1-22.

Diamandouros, N. (1983), 'Greek Political Culture in Transition: Historical Origins, Evolution Current Trends' in Clogg, R.(ed.), *Greece in the 1980s*, New York, St. Martin's Press, pp. 43-69.

EIU–The Economist Intelligence Unit (1998), 'Quarterly Report: April 24, 1998', *Country Report: Greece 1998-99*, London, EIU.

Herzfeld, M. (1985), *The Poetics of Manhood: Contest and Identity in a Cretan Mountain Village*, Princeton, New Jersey, Princeton University Press.

Konstadakopulos, D. & D. Christopoulos (1997), 'Innovative Milieux and Networks, Technological Change and Learning in European Regions', presented at the international conference 'Technology Policy and Research on Development Systems in Europe', United Nations University, Institute for New Technologies, Seville, October 1997, mimeo., 28 pp.

Lyrintzis, C. (1984), 'Political Parties in Post Junta Greece: A Case of "Bureaucratic Clientelism"?' *West European Politics*, 7 (2), 98-118.

Ritaine, E. (1998), 'The Political Capacity of Southern European Regions' in Le Gales, P. & C. Lequesne (eds), *Regions in Europe*, London, Routledge, pp. 76-88.

Verney, S. (1994), 'Central State-Local Government Relations', in Kazakos & Ioakimidis, *Greece and EC Membership Evaluated*, London, Pinter, pp.166-180.

Regionalism in Iberia

Jesús del Río Luelmo and Allan Williams

Introduction

The inclusion of Spain and Portugal together in the section dedicated to Iberia underlines the enormous differences which separate these two states which are so closely related in geographical and historical terms. The two Iberian neighbours have had sharply divergent experiences of regionalism and nationalism. Spain today has a Constitution which enshrines the rights of its regions (or, more accurately, of its nations). It also has experienced, and continues to experience, substantial democratic and violent challenges, in some regions, to the form and the very existence of the state. The Basque Country offers a dramatic example in this context, whereas in Catalonia the debate is of an exclusively political nature. These developments are themselves the culmination of four major phases of separatist movements since the beginning of the 19th century (Newton 1983, p. 98) – the *Junta* movement during the Peninsular War; *Cantonalismo* in the 1870s after the overthrow of Isabella II; nationalist movements in the early 20th century following the breakdown of the Restoration Regime, movements deeply rooted in the development and evolution of the nationalist ideologies of late 19th century; and the democratic Republican movement in the 1930s. All these phases are the recent manifestation of deeply- rooted historical differences.

In contrast, Portugal is one of the most unified nation states in Europe. Its borders have hardly changed over the last 850 years (Pereira, 1995). There are some minor exceptions, like the town of Olivenza, which is situated nowadays in Spain, but these are insignificant in terms of Portuguese territorial evolution. The presence both of the Atlantic coastline and of more or less clearly defined natural land boundaries has contributed to this fact. And, apart from some weak separatist movements in the Atlantic islands, the legitimacy of the Portuguese central authority has not been seriously questioned by any of its regions. In the remainder of this opening section, the bases, or the lack of these, for regionalism and nationalism in Iberia will be considered.

An analysis of regionalism in Portugal reveals immediate contrasts in comparison to the Spanish experience. Although, unlike Spain, there are no substantial linguistic differences within Portugal, there has been a pattern of uneven development, with economic growth being concentrated in the coastal zone between Setubal and Braga. Geographical factors were obviously at the root of this pattern. Most of the rest of the country remained relatively underdeveloped until comparatively recently, and even today these differences remain substantial. This is to a large extent the case with Spain too. Portugal, like Spain, has also been subject to a prolonged period of centralist

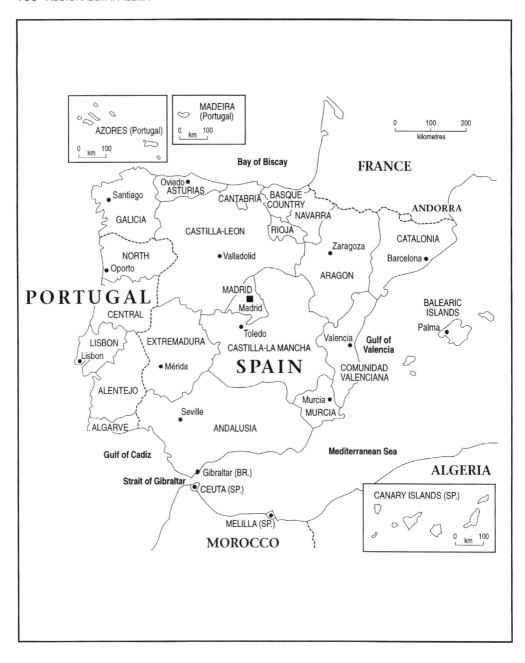

dictatorial government (1926-74). Yet there have been no regionalist or nationalist movements of any note. It is not that there are no regional differences between the littoral and the interior, between the North and the South, and between the mainland and the islands (Lewis & Williams, 1981). In this respect, it can be argued that Portugal's Alentejo has more in common with Spain's Andalusia, and the Minho has more in common with Galicia, than either has with each other. The reasons why these socio-economic contrasts have not led to regionalist movements are to be found in the processes of state formation, a shared language and culture, and the role of Lisbon in the life of the country, especially of its élites. And, not least, in the relatively minor nature of internal differences.

State Formation in Spain and Portugal

The key factor in the history of state formation in Portugal is that the state came into existence before the nation, as a result of military reconquest from Arab occupation; this greatly facilitated centralisation (Valente, 1983). This meant that, unlike in Spain, the construction of the state was due solely to conquest and not to alliances between states (Feijo, 1989). The boundaries of the state had been established by the first half of the 14th century, and the nation was formed within these. There followed another period of conquest in the 15th and the 16th centuries, when 'the adventure of discoveries and colonial expansion stressed the atlanticisation of the settlement pattern and contributed to the unity of a country turned towards the sea' (Gaspar, 1984, p. 3). This maritime calling was influenced to a large degree by the occasionally uncomfortable presence of the Spanish neighbour.

Within these boundaries a nation with a single shared language was formed. Moreover, by 1500, with a few exceptions, Portuguese had evolved to become a language clearly separate from Galician, to which it remains closely related. The importance of this close association of a distinctive language with a single unified country was immense, a point that is underlined by comparison with Spain. Language has thus proved to be one of the main factors at the root of regionalist and separatist movements in Iberia.

There is a number of other reasons for the strong unity of state and nation in Portugal, including the absence of significant minorities and the existence of a shared religion; these served to diminish although not to eliminate regional cultural differences. The unchallenged primacy of Lisbon in the urban hierarchy, and strong levels of inter-regional migration, also served to strengthen national identity and a certain cultural homogeneity. Of even greater importance, and here again there is a contrast with Spain, was the early emergence of a dominant class, the nobility, which assumed the role of national leadership and the formation of a sense of solidarity. Later this role passed to the urban bourgeoisie. Pinha-Cabral (1989, p. 15) writes that 'it is the existence of solidly entrenched urban political and economic élites that explains the apparent unquestioned continuity of Portuguese identity'. The migration of these élites to Lisbon and their education at Coimbra or Lisbon universities further underlined this hegemony. And it has made it difficult to create separate strong regional identities opposed to the mainstream centralist mentality.

Virtually the only instance of separatism in modern Portuguese history is to be found in the Azores islands. Even here there is no sustained history of separatism (Coutinho, 1978), but rather a minor upsurge in the 1970s, related especially to the drift in Portuguese politics to the far left during the course of the unpredictable transition to parliamentary democracy between 1974 and 1976 (Gallagher, 1979). There are two grounds on which a regionalist movement could be mobilised; first, remoteness from the centres of economic and political power on the mainland, and, second, the prevalence of low incomes, which are only half the Portuguese average (Williams, 1994). The fact that these grounds have never provided a basis for more that a rudimentary regionalist movement can perhaps be explained by four features of the islands' relationships with the mainland (Lewis & Williams, 1994). First, prior to their colonisation these were uninhabited islands, so their economies and societies have never known a historical evolution independent of the mainland's. Second, the education of the islands' élites has cemented their links with national élites. Third, the distances between the nine occupied islands in the archipelago, and differences in their economic and social structures, militate against a united Azorean movement. Finally, while there are differences in folk culture between the islands and the mainland, these do not extend to high culture.

Turning to Spain, it can be said that in the latter years of the Francoist regime, regionalism and nationalism became intertwined with the democratic opposition movements, and this contributed to the priority given to the regional question in the post-Franco political settlement, the so-called *Pactos de la Moncloa*. A common view amongst the analysts and a common perception in today's Spanish society is, in this sense, that despite, or perhaps because of, the 1978 Constitution which recognised the regional differences and their political materialisation, national and regional diversity is 'without any doubt the greatest political problem which the still young Spanish democracy faces today' (Guerra 1989, p. 19). Almost ten years after this statement, with a considerably more mature democracy, it can still be considered entirely valid. In Portugal, in contrast, the post-1974 democratic transition was in many ways far more dramatic, with tanks on the streets and attempted counter-coups. However, it was also unconstrained by the dictates of nationalism and regionalism.

Spain has a long and relatively early history of state formation. During the reconquest from the Arabs a number of independent Christian kingdoms were formed which were unified after the marriage of Isabella and Ferdinand in the 15th century. However, centralisation was limited, for outside Castille traditional rights, privileges and laws (*Fueros*) were respected. As a result, Spain could best be described at that time as 'a confederation of political units' (Tamames & Clegg, 1984, p. 32). The state came into existence, as in Portugal, before the nation – before the Spanish nation, but significantly after the birth of many nations which nowadays are within Spain (Maldonado Gago, 1995). This is an interesting circumstance, since it has made the process of centralisation much more difficult and has contributed to the maintenance of internal differences.

The path towards state unity was by no means easy; it even included the temporary annexation of Portugal (in the 16th and 17th centuries). Significant among the clashes between the central power and the peripheral regions was the rebellion of the Catalans in the 1640s against the Spanish monarchy, simultaneous with the rebellion of the Portuguese. This rebellion (unlike the Portuguese one) was crushed by the Spanish monarchy. The impact of this conflict can be felt in Catalan nationalism – even nowadays the song *Els Segadors*, the symbol of that rebellion, is the official anthem of Catalonia.

During the 18th century, under the Bourbons, these traditional rights were reduced but, even so, the Basque country retained its ancient rights, the *Fueros*, until the 20th century. Political centralisation did not, however, lead to a unified nation state. Regionalist pressures re-emerged in the late 19th century, partly due to the impact of uneven development, which widened the differences between the economic structures and policy requirements of regions, as well as leading to immigration which challenged the indigenous cultures and traditions of two of the historic nationalities, the Basque Country and Catalonia. In fact, in the Basque case, the re-emergence, if not the birth, of nationalism has to be analysed in direct relation to the wave of migrants from other regions during industrialisation in the 19th century. This re-emergence of regional awareness led to a limited devolution to Catalonia in the early 20th century.

In the 1930s the Republican government granted regional autonomy to Catalonia and, belatedly, to the Basque country and Galicia. The levels of autonomy reached at that time by Basques, Catalans and Galicians were without precedent in Spain's recent history. This served to renew further the cause of greater autonomy for these three historic nationalities. But the remarkable dynamism of 1930s Spanish political life was to receive a dramatic setback with the outbreak of the Civil War that gripped the country.

The victory of the so-called *nacional* (sic) side in the war left no doubts about the treatment of the question of devolution to the regions. Under Franco all vestiges of regionalism were swept away, while there was also a sustained attack on virtually all aspects of regional distinctiveness. However, Francoism failed to eradicate regionalism and nationalism. Despite these efforts, at the time of Franco's death, not only was there a long history of demands for regional autonomy in Spain, but there were large sections of the population which had actually experienced regional autonomy during their own lifetime. This, in itself, although important, is not sufficient to explain the extraordinary outpouring of regionalist and nationalist demands in post-Franco Spain. The reaction against Francoist repression is another explanation – many opposition political parties accepted and supported nationalist claims as part of a wider opposition movement and this has reinforced their role in the period following the end of the dictatorship. To understand this phenomenon, it is necessary to examine three elements – the strength of cultural and linguistic diversity, the impact of uneven development, and the nature of the opposition to Franco's regime.

Cultural differences form the roots of Spanish nationalism. This is illustrated by the way that the growth of political nationalism in the late 19th century was preceded by a cultural revival. In turn, it is clear that language lies at the very heart of cultural differences in the case of the three historic national minorities, the Catalans, the Basques and the Galicians. Terradas (1989, p. 115) emphasizes this in the case of the

Catalans, arguing that their cultural forms are 'not by themselves different from the cultural forms of other nations, it is their association with language that makes them different'. The Catalans provide only one example and the sheer scale of linguistic diversity in Spain is impressive – an estimated one quarter of the population of Spain speak a language different from Castillian. In addition to Catalan, other official languages are Valencian, Basque (which is also official in Navarre), Mallorcan and Galician.

The most distinctive language is the Basque Country's *Euskera*, a non Indo-European language, the origin of which remains a mystery. It is a minority language within the Basque Country, spoken by some 300-350 thousand persons (Williams, 1994). Even in its heartland, Guipúzcoa, there are fewer Basque speakers than there are Castillian speakers. The system of *Ikastolas* (Basque schools) and the programmes implemented by the regional government to boost the use of Basque language have achieved only a limited increase in this figure. The numbers, however, belie the symbolic importance of the language, for 'the singularity of *Euskera* encouraged the belief that the Basques were a race apart' (Sullivan, 1988, p. 2), with a right to be independent of both Spain and France. Galician and Catalan are less differentiated from Castillian than Basque (as a consequence of their common Latin roots), but both are clearly identified as separate languages, and play crucial roles in nationalist politics. In the case of Galicia, Díaz López (1982, p. 404) writes that 'the language aside from being the most radical and representative element of Galician culture, is what conferred on it the character of a differentiated people with a right to political autonomy'. Language and culture were and are also at the very heart of Catalan nationalism. Their role in the revival of nationalism in the late 19th century has already been noted, and it was language which was again the immediate focus for much of the opposition to Francoism. Franco brutally suppressed the use of Catalan in most areas of non-domestic life after the Civil War. Conditions were relaxed in the 1960s and 1970s but mostly at the level of everyday use. Catalan was not recognised as an official language or at the level of high culture, so that at best it faced the prospect of becoming a rustic or folk curiosity. This was to be one of the rallying points against Catalan opposition groups in the 1960s (Keating 1988, p. 213).

The language issue has necessarily been simplified in this discussion. Catalan, for example, is spoken outside of Catalonia, in Valencia, the Balearic Islands and some areas of Aragon, (there is in fact a fierce debate as to whether Valencian should be considered as a dialect of Catalan or as a language in its own right) while in parts of the region, such as the Val d'Aran, it is not the dominant language. Castillian is also not a simple unified language for there are important dialects in Andalusia, Asturias, León and the Canary Islands. In turn, each of these regions is socio-culturally diverse. In the case of Andalusia, for example, this is because the region has 'never had a single, centrifugal city like Barcelona or Bilbao to provide a hegemonic cultural epicentre' (Gilmore, 1981, p. 59). The result of this linguistic and cultural diversity is that Spain is a multipolar society (Guerra, 1989, p. 59), and regionalism and nationalism have flourished around these poles (López Aranguren, 1994).

While the central role of language and culture is unquestionable, these factors acquired greater potency because of the pressures resulting from uneven economic development. When industrialisation took root in the 19th century, it was largely concentrated in three regions – Asturias, and especially the Basque Country and Catalonia. The latter two were precisely the regions where the sense of cultural separateness was greatest. In contrast, the political core of the country, Madrid, remained relatively underdeveloped. Quite apart from the perception of Madrid as being a hindrance to the development of the more dynamic regions, there was also potential for conflicts over external policies. This was exemplified by the 1898 Spanish-American War. There was little enthusiasm for this conflict in Catalonia. Yet it was Catalan industry which was to suffer from the resultant loss of protected markets when Spanish colonies in the Caribbean and the Pacific secured their independence from the faltering imperial power.

Massively uneven development in the 1950s and the 1960s also contributed to the revival of nationalism and regionalism. During the years of 'the Spanish miracle', growth was again concentrated in the three traditional industrial regions, together with Madrid. Later, growth spread to regions such as Valencia, the Balearic Islands and Aragon. However, large parts of western, central and southern Spain remained relatively underdeveloped and figured amongst the poorest regions in Europe. This had three effects. First, it fostered resentment in Catalonia and the Basque Country that they had to subsidize the poorer regions. Second, in the longer term, it meant the existence of a large rump of economically disadvantaged regions which would pose difficult questions of equity and territorial justice when it came to creating regional autonomies in the 1980s. Third, and more immediately, it led to large-scale migration from the poorer regions of Spain to the more prosperous ones. In particular, it brought large numbers of Castillian-speakers into the Basque Country and Catalonia. Although this had been happening since the 19th century, the size of migration movements was of enormous importance in this period. Combined with Francoist restrictions on the use of non-Castillian languages, this led to real fears that these regions could be culturally overwhelmed. The large proportion of immigrants in these regions posed difficulties for the nationalist movements, which tried to represent both the repressed masses as a whole against Francoism and the distinctive cultures of their native peoples. This was to prove particularly difficult for the Basque separatist groups and led to consistent and almost endemic splits in both the terrorist and the political arenas of Basque nationalism (Sullivan, 1988). Nevertheless, it was the cultural challenge accompanying the arrival of the immigrant population that was to give a sense of urgency to the nationalist revivals in the 1960s.

The third key element in the revival was that regionalism and nationalism became associated with resistance to centralism and, indeed, the very existence, of Franco's regime. There had been an earlier precedent for this in the late 19th century, when attempts by the Liberals to curb the traditional rights of the regions had driven the Basques and the Catalans to the Carlist cause. After the end of the Civil War, Franco imposed strong central and political control over Spain, coupled with the persecution of non-Castillian languages and culture. When the opposition to Francoism began to re-emerge after the 1940s this was very much centred on Catalonia and the Basque

Country where the twin processes of resistance to autocracy and centralism were mutually reinforcing. This would also have longer term implications for the transition to democracy, as Keating (1988, p. 108) has emphasized: 'Franco's persecution certainly crushed the Basque and Catalan movements but, in doing so, created an enhanced sense of anti-regime solidarity in those regions and ensured that any return to democracy would need to have a regional dimension'.

While the discussion so far has emphasised the historical bases for the presence or lack of regionalist tendencies in the two countries, much of the character of regionalism and the detailed forms of regional government have been shaped during the two decades since the end of the dictatorships. These are considered in the following sections, starting with the experience of Portugal.

Regionalism in Portugal

The Portuguese administrative system is largely based on successive waves of reforms in the 19th century. After the victory of liberalism, the Napoleonic system was the model followed in Portugal, as in many other southern European countries. Reforms since then have reflected to a great extent either the absolutist or the liberal approach to administration (Pereira, 1995), but without affecting fundamentally the initial model. The present system is to a great extent the one adopted after the Liberal Revolution of 1820. Subsequent reforms had little significant effect. The abolition of the monarchy and the inauguration of the republic (1910) had only a limited impact on this territorial and administrative model. Perhaps the main legacy of Portugal's historical development has been the inertia of a unitary and centralist administration, and the almost total absence of regional movements may be linked to the fact that Portugal has one of the longer histories of statehood among European nations. Despite the undoubted importance of the roots and traditions of the Portuguese administrative system, it is the last two decades which are crucial for an understanding of the strength and nature of regionalism in the country.

Portugal during Salazar's dictatorship had been a highly centralised country. The economy had been run on corporatist lines with even minor investment and other decisions having to be referred to Lisbon. Local government financial autonomy had also been virtually eliminated as part of the regime's strategy for controlling public expenditure. It was inevitable therefore that the democratic transition would have to address the issue of decentralisation. However, the lack of regionalist issues and movements in Portugal, compared with Spain, meant that this was given a relatively low priority.

The Portuguese revolution of 1974 brought a major shift in the direction traditionally followed by both political and administrative forces. In the wake of the revolutionary events priority was given, through the new Constitution (1976), to three different processes. First to the process of decolonisation, which was immediately implemented. Second to the process of progressive democratisation, which implied the creation of a new political and electoral framework, a process now complete. And finally to the process of deconcentration. The very presence of the latter in the Constitution has a highly symbolic value, but to date it is still far from being implemented, and decentralisation is nothing but a mid- to long-term possibility.

The Portuguese political and administrative system created by the Constitution is therefore characterised by a combination of features that are peculiar to it and that differentiate it markedly from other European countries (Lynch, 1996). Jacobinism, first, played an essential role, obviously implying a strong state centralism in all aspects of political and administrative life, including the bureaucracy. This has been significant in limiting the development of a Portuguese regionalist tradition, as has the right enjoyed by locally-elected representatives and bureaucrats to gain access to Government and Parliament. The country is divided into 18 administrative districts, with some, and only some, regional significance, whose head is a representative of the central government with responsibility for certain civil and administrative services. At the local level, municipalities and local councils have a fairly strong executive, with a president at the head of each. This has no doubt contributed to a weakening of regionalist claims. Finally, the diversity of parties and an electoral system based on proportional representation are other contributory factors.

The 1976 Constitution provided for the creation of two types of regional administrative entities – first, the so-called 'autonomous regions' (the islands) and, second, the remaining territories under the general title of 'administrative regions'. The existence of a third category, in the large urban areas, was also envisaged by the Constitution, albeit in scant detail. This category came into being in 1991, with the creation of the 'metropolitan areas' of Lisbon and Oporto.

The scope provided to the regions by the Constitution is not at all clear. Three main spheres of competence are mentioned – administration of public services, coordination of municipalities and elaboration of regional plans, in addition to collaboration in the national planning process (Pereira, 1995). However, regional policies are still to a large extent managed by central agencies, and this lack of a regional tier of responsibility has further slowed the regionalisation process. A valid alternative seems to be offered by the Commissions of Regional Coordination (*Comissoes de Coordenação Regional*). Created in the 1960s and early 1970s and based in five regional areas (North, Centre, Lisbon and Tagus Valley, Alentejo and Algarve), they began life as study commissions. Their development in the 1980s has given them a political weight that could enable them, in the medium term, to take on the role of real autonomous regions, filling the gap left by the absence of proper regional governments.

The islands' situation is unique for obvious geographical reasons, which have led to the development of certain differences from the mainland, together with a more open regional awareness. Both the Azores and Madeira have achieved deeper levels of self-determination and more advanced forms of devolution than the rest of Portuguese territory, through real regional governments. Their higher degree of autonomy, compared to other Portuguese regions, was established immediately after the approval of the 1976 Constitution. This surprisingly high degree of autonomy can be explained by both 'the particular climate for democratic change in Portugal' and 'the reaction to the centralism that was associated with Salazarism' (Williams, 1994, p. 95).

The case of the metropolitan areas of Lisbon and Oporto requires a more detailed analysis. It is important to note that it involved the appearance of completely new regional entities, based on functional and practical criteria, rather than on the usual historical and geographical factors. This has been a completely new concept in Portuguese terms, as it was in the case of the creation of Madrid as an autonomous community in Spain. In the context of deconcentration, these metropolitan areas have enormous symbolic value, thanks to their supramunicipal scope. Their creation has meant, first, a break with the traditional division into regions according to more or less natural criteria. Furthermore, it creates an interesting precedent and a model of decentralisation for the rest of the Portuguese territory. It is, however, an idea with numerous limitations – neither Oporto nor Lisbon are real regions in the traditional sense of the term and their example is unlikely to be followed by other areas in Portugal. It would seem, therefore, that the likelihood of this example being followed elsewhere is more theoretical than actual.

Despite the progress described above, it is clear that deconcentration is far from being implemented and the pace of change appears very slow. The reasons for this are complex. Pereira (1995, pp. 275-279) offers a variety of explanations. First, the inertia of the centralist administrative system has played a remarkable role. Internal political rivalries have also been significant in preventing the deconcentration process from being fully implemented. The power of the Communists in the municipalities has made the national ruling party very aware of the danger that an increased share of power for local authorities, including the regions, would present for its own survival. The government's opposition is therefore not very surprising, and has tended to slow the process of regionalisation. Even the progress made to date can be seen in a negative light. The question is 'whether deconcentration is a driving force for or a restraining power against regionalisation' (Pereira 1995, p. 278). A process of deconcentration as it is conceived in Portugal can be considered as a rational way of improving public services and allowing the regions to gain progressively greater influence. For the supporters of devolution, however, this is seen as a manoeuvre to postpone regionalisation, and any improvement in services will therefore imply, in this light, a strengthening of the role of State administration in the regions.

Regionalism in Spain

Regionalist awareness and devolution are by no means new phenomena in Spain. As we have already seen, the modern state is formed out of the union of several different medieval entities, the majority of them independent kingdoms, which came together under the rule of Isabella and Ferdinand in late 15th century. There is therefore a long common history, around 500 years without major changes in either borders or internal composition, but with substantial differences underlying this apparent unity. The complexity of the Spanish situation, resulting from a mixture of historical, cultural, economic, ethnic and political factors, implies that any analysis of regionalism in Spain carries the risk of including in the same term quite different categories. These range from a diffuse regional awareness in some of the recently-created autonomous communities, to the deeply-rooted independentist feeling of the historic nations. The collapse of the Francoist dictatorship brought these differences to the fore and laid the foundations for the process of devolution.

According to Lausen (1986) Spain resembles an inverted centre-periphery model. The principal feature of this model is that the prime instinct of the centre is to retain power rather than to promote economic integration and growth. The result is an unstable territorial system in which there is oscillation between repressive centralism and unstable decentralisation. By the latter years of the Franco regime, repressive centralism, together with the social tensions related to uneven development and the lack of democratic channels for regional representation, has brought about a regional backlash of varying intensity throughout Spain. The most extreme expression of this was to be found in the Basque Country where ETA, after its Fifth Assembly in 1966-7, launched an escalating campaign of violence against the regime, culminating in the assassination of Prime Minister Admiral Carrero Blanco in 1973. Given the political construction of Francoism, the state's only response to this was increased repression and brutality, which had the effect of mobilizing further support for ETA, besides attracting sympathy from more moderate sectors. This was particularly marked in the regime's handling of the Burgos trial in 1970, which led to strong reactions both in the Basque Country and in the world's press (Sullivan, 1988). While responses in Catalonia and Galicia were far less violent, regional feeling, in Catalonia at least, was no less strong (Rubiralta Casas, 1997).

The tensions in the regime, and the close association of Francoism with centralism, meant that after 1975 'a solution to the regional problem was considered top priority for the new regime. In the early years of the transition to democracy it was felt that these tensions could endanger the democratic process and even the territorial integrity of Spain' (Guerra, 1989, p. 21). While decentralisation was top of the agenda, there was little agreement as to the form that this should take. A regionalist model which made special provisions for the historic nationalities, the Basques the Catalans and the Galicians, offered the quickest and simplest solution. But this had two flaws – it would give political privileges to what were already economically-favoured regions (except for Galicia); and its very exceptionalism would irritate right-wing hard-liners in the armed forces and other areas of the state apparatus. The alternative was the broad sweep of federalism, offering a similar package of decentralizing measures to all regions. This model also had its drawbacks, not least in that there was no agreement in all parts of Spain as to what constituted regions. Regional divisions were extremely controversial at the time of the creation of the autonomous communities. The differences remain nowadays in certain areas, with varying degrees of intensity – the fusion of Castilla and Leon, the somewhat artificial creation of Castilla-la-Mancha, the inclusion of Murcia and Cartagena in the same province, and particularly the separation of Navarra and the Basque Country, strongly contested and defended by the political forces of both communities, are some examples (Alonso Fernández, 1993). In general, however, these can be seen as minor problems, and the overall pattern enjoys a wide consensus. Inevitably, the implementation of any federalist solution was felt to be too fast for some regions and too slow for others, such as Catalonia and the Basque Country.

In their manifestos for the 1977 election, all the major parties, except the *Alianza Popular* (later *Partido Popular*, which favoured administrative decentralisation), committed themselves to a version of one of these two models. The victorious *Unión de Centro Democrático* (UCD) party favoured a flexible formula whereby the constitution guaranteed the right to autonomy but the initiative for autonomy was given to the regions themselves (Newton, 1983, pp. 105-106). This was reflected in the 1978 Constitution which, inevitably, was a compromise between the two models; it advanced a regionalist model in the short term and a federalist model for the longer term (Bernecker, 1990). It was a concoction that Pi-Sunyer (1988, p. 2) has fittingly called *café para todos*.

The 1978 Constitution is based on three main principles – the unity of the state, regional autonomy and interregional solidarity. Formally, Article 2 states that 'the Constitution is based on the indissoluble unity of the Spanish nation, the common and indivisible motherland of all Spaniards, and recognizes and guarantees the right to autonomy of the nationalities and regions of which it is composed and the common links that bind them together'. Each region was given the opportunity to organize itself into an autonomous community and to draft its own statute of autonomy which would then be submitted to the parliament for approval. In practice, three main routes were laid down for achieving autonomy (Alonso Fernández, 1993) – first, the regions with a historic claim to autonomy were offered relatively higher levels of autonomy. These were defined as the regions which during the Second Republic had voted for autonomy in a referendum. As such, it included the three historic nationalities: Catalonia, the Basque Country and Galicia. This does not mean that other regions excluded from this category do not have a long history (Castille-Leon is a good example); second, a provision was made for other regions to argue that they had exceptional cases for higher level of autonomy. Eventually, four regions, Andalusia, Valencia, Navarra and the Canaries, were able to avail themselves of this route; and third, the remaining regions were offered lower levels of autonomy during the first five years, but could then seek to revise and to increase their autonomous powers subject to a maximum ceiling level. Ten regions were to take this route and all were to revise their autonomy statutes after the first five years.

It was inevitable that any model would generate tensions between Madrid and the regions as to the extent of regional autonomy. However, the very flexibility of the model exacerbated the conflict; the historic regions felt that they were insufficiently privileged and differentiated from the other regions, which in turn considered that they were being discriminated against compared to Catalonia, the Basque Country and Galicia. Given that the Constitution allowed for negotiation of the range and level of powers to be decentralized from Madrid, the end result was an enormous and confusing variety of autonomous statutes.

Implementation of the Constitution has been slow and beset with difficulties, and is still not complete. Catalonia and the Basque Country were successful in getting their statutes approved relatively quickly in 1979; these provided for significant autonomous powers, including the rights to establish autonomous education systems, separate police forces, and independent (of Madrid) television networks. The Basque and

Catalan languages also acquired official status alongside Castillian. These measures were particularly important in meeting the demands of the Catalans for cultural autonomy. Thereafter, the conservative elements within the UCD government sought to reduce the level of autonomy offered to the other regions, and to delay the process of regionalisation. As a result, Galicia's statute was not finally approved until 1981. Andalusia was next in line and the government now sought to increase the qualifying hurdles for regions seeking autonomy. In the referendum in Andalusia to approve the establishment of an autonomous region, there was an overall majority in favour, but there was a failure to obtain 50 per cent of the votes cast in just one province. Under the new rules for the referenda, this meant that the motion had failed, and the government seemed to have halted the swelling tide of regionalism. However, this proved to be 'a pyrrhic victory' (Newton, 1982, p. 30). There was a strong backlash in the region against the undemocratic procedures for the referenda, and this actually served to mobilise support for the cause of autonomy. Eventually, a second referendum was held on the revised basis of simple majority voting in the region as a whole. The second ballot produced an overwhelming vote in favour of autonomy for Andalusia.

In 1981 there was a dramatic turn of events when there was an attempted coup. Given the UCD government's and later the Socialist governments' fear of being outflanked by the right, this added to the pressures to slow down the reforms. The 23 February events meant if not a threat to democracy, at least a serious warning to the pace the reforms were being implemented. Clark (1987, p. 140) has a more cynical view of the events and believes that the national parties in the late 1970s had courted the support of the regionalist and nationalist parties while they needed their support in the democratisation process, but by the 1980s 'Spain's élites see no need to continue the transfer of power to the Autonomous Communities'. This association between national opposition parties and regionalist and nationalist ones is one of the factors at the root of the favourable treatment given to the regionalist question in the early years of Spanish democracy.

Which of these explanations is more accurate is a matter of debate. However, the outcome was decisive – the UCD government, with the support of the Socialist opposition, introduced in 1982 the *Ley Orgánica de Armonización del Proceso Autonómico* to harmonize and limit Spain's regionalism programe. It ordained that all other regions were to take the slow route to autonomy and, retrospectively, reduced the powers that had already been granted to the Basque Country and Catalonia. However, this was not the end of the saga; in 1983 the Constitutional Court ruled that large parts of the new law were unconstitutional (Pi-Sunyer, 1988, p. 8). Thereafter, there was a rapid approval of all the remaining statutes of autonomy, and regional elections were held in May 1983. As stated previously, ten regions had to take the slow route to autonomy, being granted relatively limited powers until 1989. Thereafter, all the regions had 'full' autonomy although the actual distribution of powers varied from region to region. Despite some limitations in, and some disappointments over, the final territorial settlement, this still represented a major advance for the regions, one that would have seemed unthinkable in the early 1970s.

Given the expectations surrounding this programme of regionalism, especially amongst the historic nationalities, and the inauspicious economic climate for any territorial settlement in the 1970s and 1980s (Rhodes & Wright, 1987, p. 15), not to mention the exigencies of the Spanish democratic transition, it was inevitable that the reforms would fail to satisfy all national and regional interest groups. In practice, the process of devolution has often been muddled and indecisive, but by the late 1980s major reforms had been implemented, which were probably unparalleled, in terms of speed and scale, elsewhere in western Europe. The 1978 Constitution opted for a model of diversity within unity (although the latter seems continuously under threat from nationalist pressures). Several reasons explain the creation of this constitutional model. Nationalist movements had won a great deal of support during the last years of the dictatorship. They were supported by many opposition forces at that time, and this support was maintained after Franco's death – they perceived a community of interests, something that the post-Franco experience has proved to be wrong. But the symbolic value of the nationalist movements' opposition to the former regime was to a great extent taken into account by the Constitution. The regime had tried to suppress them violently, and its collapse was seen by both nationalists and opposition political parties as the right time for the establishment of a new order headed by these symbolic elements. All these circumstances led to the need for the creation of a national consensus after Franco's death. Transition to a democratic political system was not easy, and the nationalist movements had to be taken into account (Bosch Gimpera, 1996). The need to achieve an equitable balance allowed these regions to gain a rapid devolution of power which, in other circumstances, would have been much slower and more difficult. The creation of a constitutional framework gave a voice to the claims of these groups. The democratic system allowed both respect and encouragement for these regional (and national) identities, something which would have been completely unthinkable under the former regime. That this treatment was applied to all of them and not merely those with a stronger identity can be attributed to the need for equity. It also formed part of the search for a balanced model of territorial administration which was both different from the Francoist model, and efficient (Mar-Molinero & Smith, 1996).

The result of this devolution process has been enormously complex. Spanish territory has been divided into 17 autonomous communities (*Comunidades Autónomas*), with six official different languages (Spanish or Castillian, Basque, Catalan, Galician, Valencian and Mallorcan). This itself is extremely important if seen in conjunction with the process of devolution that has yet to reach its conclusion. This complexity goes beyond the division established by the 1978 Constitution. Drawing the line between regionalism and nationalism is extremely difficult in some cases. The dynamism with which the process has developed is remarkable, particularly when we analyse the growing regional awareness of some of the recently-created entities. The situation remains fluid, and conflicts over competencies, parliamentary alliances and clashes over a variety of political and economic measures are the main fuelling forces of this development at present.

There has been a real decentralisation of power so that in a very short time even the smallest regions have become responsible for local roads, agriculture, planning economic development and the collection of indirect economic taxes. Regional governments, set up by direct election, have been consistently gaining new powers. In the case of the historic nationalities, and some of the other larger regions, they have become responsible for almost all state functions within their boundaries, except the police, the army, foreign policy, major infrastructure and some of the nationalized industries. This has created far greater scope for regional economic policy initiatives. In practice, most regional governments have been relatively conservative in this arena, even if more responsive to regional needs. One of the few radical experiments was the attempt by the Andalusian regional government to introduce land reforms so as to expropriate and redistribute the lands of underused latifundist estates; this was blocked by the central government and the Constitutional Court. Instead, most of the regions have directed their attention to education and culture. In Catalonia the reforms have enabled equal status to be given to Catalan alongside Castillian, and Catalan television and radio stations have been established. While this has generated new conflicts with the Castillian-speaking population, immigrants and locals alike, in these regions, it has also met some of the immediate demands of the historic nationalities.

One of the key tests in any regional autonomy settlement is the rearrangement of financial powers between the centre and the regions. In this respect, the reforms have been decidedly conservative. The Basque Country and Catalonia have been given special rights because of their historic *fueros* and are empowered to collect all taxes within their territories, except customs, petrol and tobacco levies, even though most of the tax rates are dictated by Madrid. Conflict over the level of these taxes is part of the daily agenda in Spain. The national Supreme Court overruled some of the tax policies of one of the Basque provinces in March 1998, provoking a new clash between central and regional governments, and creating concern over the limits of the system. However, in the case of other regions most revenue (in some cases more than 90 per cent) comes from the centre, although some degree of autonomy in spending is permitted (Hebbert, 1990, p. 126). In contrast, the percentage coming from their own taxes is of limited significance, although many regions supplement this with other local sources of income, including lotteries. Initially, the levels of these financial transfers from the centre were based on bilateral negotiations with the regional authorities. Later they were to be based on formula-funding involving criteria such as population and per capita incomes, so that there would be an in-built redistributive bias.

In addition, two other redistributive mechanisms were created to assist the poorer regions – special grants were made available to regions with sub-standard levels of services; and the Interterritorial Compensation Fund redistributes up to 30 per cent of all new public investment to the least-favoured regions. This fund was actually instituted in 1984 and it determined the levels of 'compensation' on the basis of four indices – per capita income, the migration balance, unemployment and surface area. In 1985 this meant that the largest transfers were made to Andalusia and to Galicia (Donaghy & Newton, 1987, p. 106). While there is a strong case for such transfers on the grounds of equity, the transparency of the mechanisms used has made them highly political. As a result, financial devolution has been, and is likely to remain, a major point of contention between the regions and central government.

The process of inter-regional transfers is critical for the future of Spain which remains a country with deep regional economic differentials. Despite the dynamism of the Spanish economy in the 1980s and the apparent recovery in the late 1990s, some of the poorest regions are to be found in Spain in, for example, Extremadura, Andalusia, Galicia and both Castilles. At the other extreme are Barcelona and Madrid, which have claims to be considered world cities. One of the ironies of *La España de las Autonomías* has been the increasing economic dominance of Madrid, which has more than one third of all corporate headquarters and has attracted a disproportionate share of foreign direct investment. As Hebbert (1990, p. 135) notes, 'whatever the centre has lost in the devolution of administrative and political power, it has regained in the concentration of corporate power'. The claims on central government resources to redress these imbalances are immense, and the very existence of the regional assemblies and governments has created potent channels for the articulation of such regional demands. As a result, in the 1993 parliamentary elections, the *Partido Popular* made strong advances in Valencia and other regions by claiming that the Socialist government had neglected them by concentrating central expenditure on projects such as the high speed Madrid-Seville rail link. Such disputes over discretionary central expenditure serve to reinforce those over formula funding.

Another major problem, and perhaps an inevitable one, has been the enormous differences between the regions in terms of regional identity. While some regions and nationalities have very strongly-developed collective identities, others such as Murcia and La Rioja, not to mention Madrid, are considered to be little more than artificial constructions designed to fill in awkward gaps in the map of the regions. There are also major differences in population size, ranging from about 260,000 in La Rioja to about seven million in Andalusia; this has meant that there are strong differences in the resources available to these regions and in the functions which it is appropriate to devolve to them. Finally, there are also differences between the large multi-province regions such as both Castilles and Andalusia, and the single province ones such as Cantabria. Castille-Leon is the best example of the former; it is larger in area than some of the European Union's Member States (Portugal, Ireland, Belgium, Netherlands, Luxemburg, Denmark), with a population of around 2.5 million. While this has caused difficulties in the short term, especially in the transfer of powers, in the longer term the very existence of these regions will probably lead to the reinforcement of the sense of regional identity in even the weakest of the new regions.

One of the major tests of the new regional system must be the extent to which it has met the demands of regional and national populations and interest groups. The foregoing discussion has made it clear that, thus far, there has been greater progress in respect of culture and education than in economic policy. However, the experience of autonomy to date has fired the desire for further devolution. In particular, the historic nationalities, and some of the other regions, are demanding a more radical redistribution of financial autonomy from the centre to the regions. Despite these reservations, the regional reforms have changed the political landscape of Spain in many significant respects. Nowhere is this clearer than in the Basque Country where the twin processes of democratisation and regionalisation have removed much of the support for violent action by ETA and other groups (Conversi, 1997). While it is true that Herri Batasuna, the political wing of ETA, still regularly attracts the support of 14-16 per cent of the electorate (Ross, 1991), this is an exception. Most Basque parties, which together attract one half to two thirds of electoral support, are in favour of the democratic process, so that violence is no longer seen as a legitimate route to further autonomy and independence.

Finally, it is necessary to take into account in this overview the fact that the presence of regionalist, or nationalist, political parties in the Spanish Parliament has had a very significant impact on the role of the regions. This was particularly visible during the most recent period of PSOE (*Partido Socialista Obrero Español*) government, when a coalition with the Catalan nationalist party (*Convergencia i Unió*) was required before they could obtain the majority they needed to govern. The same path has been followed even more recently by the conservative party (*Partido Popular*) when it found itself short of a majority after the 1996 elections. The fact that a nationalist party such as *Convergencia i Unió* has, to a large extent, dictated the pace of national politics since 1994 has undoubtedly exerted a major influence on the progress of devolution.

Conclusions, Recent Developments and Future Scenarios

It is hard to avoid the temptation to establish similarities between the two Iberian neighbours in an initial approach to the issue of regionalism. Indeed, there appear to be enough grounds to justify such comparisons. First, the two countries' geographical situation on the Iberian peninsula, on the periphery of the European land mass has created a series of common points which have clearly played a major role in their development. They share common roots and a broadly similar historical development over the centuries. More recently, political developments in both countries have followed a similar pattern, with dictatorships in both states which came to an end almost at the same time, giving way to democratic regimes. Certain common concerns and interests seem to reinforce this idea.

These similarities are, however, an illusion. Close analysis reveals immediate differences. As we have seen, social, economic and political development has been substantially different in each country, and the regional concept has therefore evolved very differently in each case. Territorially, although similarities can be established between Portugal and the neighbouring territories of Spain, the reality of the Spanish situation is much more complex. The size of Spain, its territorial variety and the presence of several languages make comparison with Portugal intrinsically difficult. All these factors have created a more centralist tendency in the Portuguese case, while centrifugal tendencies seem to characterise Spain both historically and in the present. Any conclusions must therefore treat the two cases separately.

Portuguese devolution, if such it can be called, is still embryonic and there seems little prospect of dramatic developments in present circumstances, given the absence of significant differences in the ethnic composition of the population, minority languages or levels of development. The creation of the metropolitan areas of Oporto and Lisbon, together with the initiatives taken by the Regional Coordination Commissions, offer only limited prospects of change.

While the influence of Europe has, as we have seen, aroused a degree of regionalist sentiment, it has not been enough in itself to create a genuine regional conciousness. In fact, in both Spain and Portugal the process of regionalisation started, with varying degrees of intensity, success and motivation, before the integration of either country in the EC, but largely independently of this event. One of the main effects of accession has been the choice of the regional scale for the distribution of the European Union's structural funds, and hence the need for the (formal) establishment of regional divisions. Although the fact that Community funds are distributed according to regional divisions has proved decisive elsewhere, the impact in Portugal has been much less obvious, simply because the whole territory comes under the Objective 1 scheme. Local development has, however, been affected, 'Since the late 1980s, the availability of European Community resources has sponsored an increase in rural development projects accross Portugal' (Syrett, 1995, p. 285) and this could lead in the medium term to a growth in regional awareness. So far, though, the influence of the European Union has been felt far more at a local, rather than at a regional, level.

In contrast, Spanish regionalism seems to have reached a point at which disagregation of the state itself is at issue. The fact that the limits of the constitutionally established model are about being reached makes conflict over continuing devolution likely (García Ferrando *et al.*, 1994). Some autonomous communities, notably the Basque Country and Catalonia, are characterised by their independentist claims. Still others are going through a process of growing nationalist awareness (Galicia, Andalusia or Valencia, for example), with an increase in the nationalist party's vote. This, of course, is likely to cause further clashes between the central and regional government ones over competencies and responsibilities. It could even bring about changes in the Constitution. Even communities which previously displayed no evident regionalist sympathies now seem to be developing regional sentiment, and in those cases where it existed in a weak form it now appears to be growing in strength. Thus devolution, which has often been perceived as benefiting only the Catalans (as they are the main nationalist force in the national parliament), has had a significant impact on the remaining autonomous communities. While this phenomenon has its positive side, it is hard to escape the conclusion that fresh conflicts will arise between regions and central government in the foreseeable future.

The so-called historic nationalities, particularly Catalonia and the Basque country, have already made substantial gains in the field of devolution. In the Basque Country, movement towards greater autonomy has been hampered by terrorist activity, and this has elicited a powerful response both within the region and (more predictably) in the rest of Spain. While such responses may have a positive impact in bringing about an end to violence, they have also provided a new element of dissension in an already very divided society. The question arises as to whether the result could be the excision of the most advanced regions (notably Catalonia and the Basque Country), and whether this would actually make any practical difference. Opinions are evenly divided, given that the model of devolution to the autonomous communities is as yet incomplete, while devolution claims continue to grow.

The impact of both the Maastricht Treaty, and subsequently the Treaty of

Amsterdam, together with the creation of the Committee of the Regions, has been relatively slight compared to the impact of internal developments in both countries. The presence of the Catalan nationalists in the governmental coalition in Spain and the creation of new regional entities in Portugal are significant elements in these developments. But the limited impact of European Union measures can be explained in other ways. The Committee of the Regions has received much attention from some regionalist groups and nationalist political parties. So far, however, it has always been perceived as part of a mid-term strategy which seeks to strengthen the role of the regions within the European Union. The creation of schemes for interregional cooperation has come as an independent initiative rather than as a part of European Union strategy (Morata, 1997).

The creation of the Committee of the Regions by the Maastricht Treaty is regarded, therefore, as an interesting idea in the medium term by the representatives of the regions, but its lack of a more precisely defined institutional framework at present makes it very fragile. In fact it can be easily overwhelmed by initiatives such as those proposed by the Spanish government in March 1998 which foresee the possibility of the participation of its regional representatives in the European Union's Council of Ministers. Significantly, neither Morata nor Pereira (1995) mention any significant impact of Maastricht in the processes related to Iberian regionalism. The former even makes the case that Spanish integration in the European Community has had a negative effect on the institutional role of the regions, but without making any direct reference to Maastricht.

Future developments are difficult to predict, although it is possible to argue that the European Union will in future have to play a greater role with both nation states and regions (Wagstaff, 1996). In the Iberian context the role played to date has been minimal. Spain, more deeply affected by both regionalist and separatist movements, seems to be evolving towards a form of federation, although this is not the aim of most of the regional governments. However the very presence of Catalan and Basque nationalisms, although untypical in many ways, complicates the picture considerably and makes any forecast a risky business. There seems little likelihood, on the other hand, that Portugal will develop strong regionalist movements in the foreseeable future. In contrast to the Spanish example, Portugal is likely to remain the model of the unified nation state in Europe.

References:

Alonso Fernández, J. (1993), *La Nueva Situación Regional*, Madrid, Editorial Síntesis.

Bernecker, W. L. (1990), 'Spain and Portugal between regime transition and stabilized democracy', *Iberian Studies*, 19, 32-56.

Bosch Gimpera, P. (1996), *El problema de las Españas*, Málaga, Algaraza.

Clark, R. P. (1987), 'The question of regional autonomy in Spain's democratic transition', in Clark, R. P. & M. H. Haltzel (eds), *Spain in the 1980s*, Cambridge, Mass., Ballinger.

Conversi, D. (1997), *The Basques, the Catalans and Spain: alternative routes to nationalist mobilisation*, London, Hurst.

Corkhill, D. (1992), 'Imperfect bipolarism? Portugal's political system after the 1991 Parliamentary election', *Journal of the Association for Contemporary Iberian Studies*, 5, 16-23.

Coutinho, A.B. (1978), *Que Futuro para os Açores?* Lisbon, Editorial Caminho.

Donaghy, P. J., & M. T. Newton (1987), *Spain: A Guide To Political and Economic Institutions*, Cambridge, Cambridge University Press.

Feijo, R. C. (1989), 'State, nation and regional diversity in Portugal: an overview', in Herr, R. & J.H.R. Polt (eds), *Iberian Identity: Essays on the Nature of Identity in Portugal and Spain*, Berkeley, Institute of International Studies.

Gallagher, T. (1976), 'Portugal's Atlantic Territories: The Separatist Challenge', *The World Today*, pp. 353-360.

Gallagher, T. (1979), 'Portugal's relations with her Atlantic territories', *The World Today*, March.

García Ferrando, M. (1982), *Regionalismo y autonomías en España, 1976-1979*, Madrid, Centro de Investigaciones Sociológicas.

García Ferrando, M., E. López Aranguren, & M. Beltrán (1994), *La conciencia nacionalista y regional en la España de las autonomías*, Madrid, Centro de Investigaciones Sociológicas.

Gaspar, J. (1984), 'The unity and individuality of Portugal', *Iberian Studies*, 13, 3-7.

— (1990), 'The new map of Portugal', in Hebbert, M. & J. C. Hansen (eds), *Unfamiliar Territory: The Reshaping of European Geography*, Aldershot, Avebury Press.

Gilmore, D. D. (1981), 'Andalusian regionalism: anthropological perspectives', *Iberian Studies*, 10, 58-67.

Guerra, L. L. (1989), 'National and regional pluralism in Contemporary Spain', in Herr, R. & J.H.R. Polt (eds), *Iberian Identity: Essays on the Nature of Identity in Portugal and Spain*, Berkeley, Institute of International Studies.

Hardy, S., M. Hart *et alii* (eds) (1995), *An Enlarged Europe. Regions in Competition?* London, Regional Studies Association

Hebbert, M. (1990), 'Spain – A Centre Periphery Transformation', in Hebbert, M. & J.C. Hansen (eds), *The Reshaping of European Geography*, Aldershot, Avebury Press.

Jones, B., & M. Keating (eds) (1995), *The European Union and the Regions*, Oxford, Clarendon Press.

Keating, M. (1988), *State and Regional Nationalism: Territorial Politics and the European State*, London, Harvester Wheatsheaf.

Keating, M., & J. Loughlin (1997), *The Political Economy of Regionalism*, London, Frank Cass.

Lausen, J.R. (1986), *El Estado Multi-Regional – España Descentrada*, Madrid, Alianza Editorial.

Lewis, J.R., & A.M. Williams (1981), 'Regional uneven development on the European periphery: the case of Portugal', *Tijdschirft voor Economische en Sociale Geografie*, 72, 81-98.

— (1984), 'Social cleavages and electoral performance: the social basis of Portuguese political parties, 1976-1983', *West European Politics*, 7 (2), 119-137.

— (1994), 'Regional autonomy and the European Communities: the view from Portugal's Atlantic islands', *Regional Politics and Policy*, 4 (2), 67-85.

López, C.E.D. (1982), 'The politicization of Galician cleavages', in Rokkan, S. & D. W. Urwin (eds), *The Politics of Territorial Identity: Studies in European Regionalism*, London, Sage.

López Aranguren, E. (1994), 'Nacionalismo, Regionalismo y Postnacionalismo en las Comunidades Autónomas del Estado Español', *Razón y Fe*, 230 (1153), 269-281.

Lynch, P. (1996), *Minority Nationalism and European Integration*, Cardiff, University of Wales Press.

Maldonado Gago, J. (1995), 'España, una nación de naciones', *Política y Sociedad*, 20, 23-33.

Mar-Molinero, C., & A. Smith (eds) (1996), *Nationalism and the nation in the Iberian peninsula: competing and conflicting identites*, Oxford, Berg.

Morata, F. (1995), 'Spanish Regions in the European Community', in Jones, B. & M. Keating (eds), *The European Union and the Regions*, pp. 114-133.

Morata, F. (1997), 'The Euro-region and the C-6 Network: The New Politics of Sub-national Cooperation in the West-Mediterranean Area', in Keating, M. & J. Loughlin (eds), *The Political Economy of Regionalism*, pp. 292-305.

Newton, M. (1982), 'Andalusia: The Long Road to Autonomy', *Journal of Area Studies*, 6, 27-32.

— (1983), 'The peoples and regions of Spain', in Bell, D. S. (ed.), *Democratic Politics in Spain*, London, Pinter.

Payne, S. (1971), 'Catalan and Basque nationalism', *Journal of Contemporary History*, 6, 15-51.

Pereira, A. (1995), 'Regionalism in Portugal', in Jones, B. & M. Keating (eds), *The European Union and the Regions*, pp. 268-280.

Pi-Sunyer, O. (1988), 'Catalan politics and Spanish democracy: an overview of a relationship', *Iberian Studies*, 17, 1-16.

Pinha-Cabral, J. de (1989), 'Sociocultural differentiation and regional identity in Portugal', in Herr, R. & J.H.R. Polt (eds), *Iberian Identity: Essays on the Nature of Identity in Portugal and Spain*, Berkeley, Institute of International Studies.

Porto, M. (1984), 'Regional development in Portugal: the institutional framework', *Iberian Studies*, 13, 7-16.

Rhodes, R.A.W., & V. Wright (1987), 'Introduction', *West European Politics*, 10, 4 (Special Issue on 'Tensions in the Territorial Politics of Western Europe'), 1-20.

Ross, C. (1991), 'Towards the Basque election of 1990: the nationalist realignment of 1986-1989 and its effects', *Journal of the Association of Contemporary Iberian Studies*, 4, 49-59.

Rubiralta Casas, F. (1997), 'Le nouveau nationalisme radical dans l'Etat espagnol: analyse comparative d'un processus univoque en Galicie, en Catalogne et au Pays Basque', *Revue International de Politique Comparée*, 4 (2), 459-472.

Sullivan, J. (1988), *ETA and Basque Nationalism: The Fight for Euskadi 1890-1986*, London, Routledge.

Syrett, S. (1995), 'Local Economic Development in Peripheral Rural Areas. The Case of Portugal', in Hardy, S., M. Hart *et alii* (eds), *An Enlarged Europe. Regions in Competition?* pp. 280-294.

Tamames, R., & T. Clegg (1984), 'Spain: regional autonomy and the democratic transition', in Hebbert, M. & H. Machin (eds), *Regionalism in France, Italy and Spain*, London, London School of Economics, International Centre for Economic and Related Disciplines.

Terradas, I. (1989), 'Catalan Identities', in Herr, R. & J.H.R.Polt (eds), *Iberian Identity: Essays on the Nature of Identity in Portugal and Spain*, Berkeley, Institute of International Studies.

Valente, V. P. (1983), *Tentar Perceber*. Lisbon, Imprensa Nacional.

Valle, T. del (1989), 'Basque ethnic identity at a time of rapid change', in Herr, R. & J.H.R.Polt (eds), *Iberian Identity: Essays on the Nature of Identity in Portugal and Spain*, Berkeley, Institute of International Studies.

Wagstaff, Peter (1996), 'Nations, Regions and the Future of Europe', *Journal of Area Studies*, 9, 126-141.

The Committee of the Regions of the European Union

Peter Wagstaff

Contributors to this volume have, by and large, been sparing in their use of the phrase 'Europe of the Regions'. This is perhaps just as well, given the range of reactions it elicits and the ambiguities it enshrines. In some quarters, it spells the promise of a Europe in which the nation state is no longer the primary unit of governance – within the putative federal state of Europe it will be the regions that form the principal sub-national level. While, as events over the last decade have made abundantly clear, it is unwise to set too much store by the permanence of boundaries and institutions, it is equally clear that there is no immediate prospect of such an outcome, not least because of the widely different strengths of regional feeling, or identification, among the citizens of the European Union. A more widely shared acceptance of the phrase 'Europe of the Regions' is that which acknowledges the fact of regional allegiance, however unevenly spread, as well as the value, in both socio-political and economic terms, of an intermediate level of responsibility and decision-making, an interface between the local and the national or supra-national.

It is in this latter context that the decision taken at Maastricht to set up the Committee of the Regions of the European Union (CoR) is significant. Indeed, the lengthy debates and negotiations which led to its inauguration themselves reveal something of the variety of motives and interests at issue. The German *Länder* and, more recently, the Belgian and Austrian federal provinces, had expressed reservations about the way in which the distribution of European Union regional support was increasingly held to be the responsibility of governments at the national level, therefore limiting their own room for manoeuvre. The pressure for some sort of regional representation at the Council of Ministers grew as a result. In a parallel development, the movement from the mid-1980s towards greater integration in the European Union as a whole, evident from the Single European Act of 1987 to the Maastricht Treaty and beyond to the Treaty of Amsterdam in 1997 has, it is argued, not only increased the complexity of inter-state relations but, by eroding the autonomy of individual Member States, brought regional factors to the forefront (Loughlin, 1997, p.148). Pressure for a regional level of representation was chiefly felt in those states where the federal principle was already firmly established. Nevertheless, there was support, in varying degrees and for a variety of reasons, from the regional components of other European Union Member States. For some, notably in Spain and, to a lesser extent Italy, the regional agenda reflected aspirations towards internal political settlements. In France, newly empowered regions in a centralised state were attracted by the idea of a direct conduit to the European institutions, particularly in view of the resource implications of a growing European Union regional policy budget, now second only to the CAP in size. Elsewhere, in centralised states such as the United Kingdom and Ireland, regions which had either atrophied through neglect or been emasculated by centralisation, began to recognise that the European Union offered opportunities for regional funding and therefore a certain legitimacy denied them by their own governments. Other factors came into play as well so that, for example, it has been in the interests of a highly centralised Greece to consider some measure of decentralisation in order to justify European Union regional funding. Scandinavian countries, with a strong local government tradition, see the advantages of both inter-and intra-state cooperation. Finally, the piecemeal growth of trans-national European associations such as the Atlantic Arc and the Mediterranean Arc testifies to a willingness to explore networks of relationships which go beyond national boundaries. The Assembly of European Regions, made up of regions from eastern Europe as well as the European Union had also, since its inception in the mid-1980s, sought to lobby in favour of greater influence for regional forces in European policy-making.

The establishment of the European Union Committee of the Regions as an outcome of the Maastricht Treaty negotiations was therefore timely, although it was overshadowed by the main agenda items – the preparations for European Monetary Union, and the debate over the concept of 'subsidiarity'. The importance of this concept to the debate over regional levels of responsibility cannot, however, be overstated. The provision in the Maastricht Treaty (Article 198a) is for 'an advisory committee of representatives of regional and local authorities, hence to be called the Committee of the Regions'.

Two apparently conflicting views of the role of the Committee of the Regions can be seen in the respective positions of the United Kingdom and Germany at the time of its inception. It seems clear that the British government wished at all costs to avoid the empowerment of a regional or sub-national level of government that could be seen as a step towards a federal Europe. By endorsing the creation of an exclusively consultative body devoid of legislative powers it could claim to be cooperating with European partners without endangering its own Member State role. The perspective of the German *Länder* was, unsurprisingly, somewhat different since, for them, the CoR represented the opportunity to put into practice on a European-wide scale experience in decentralised government that had served the Federal Republic well over more than 40 years (Collins & Jeffery, 1997, p. 6). In contrast, the federal government in Bonn may well not have been displeased to see the ambitions of its own *Länder* restrained by the reticence of the United Kingdom.

When the CoR was first constituted in 1994, it was made up of 189 members, allocated according to the size of the Member State (24 each for France, Germany, Italy and the United Kingdom; 21 for Spain; 12 for Belgium, Greece, the Netherlands and Portugal; 9 for Denmark and Ireland; 6 for Luxembourg). Following the enlargement of the European Union at the start of 1995 the number of members rose to 222 with the accession of Austria and Sweden (12 members each) and Finland (9). An aspect of the

THE 15 MEMBERS
OF THE
EUROPEAN UNION

Committee's constitution that led to vigorous debate in its early stages was the freedom accorded to Member State governments in deciding how to fill their allocated seats. The result was what has been termed 'a highly heterogeneous body' (Collins & Jeffery, 1997, p. 6), although this was in any case inevitable – to have subjected the term 'region' to a strict interpretation would have excluded nearly half the Member States (United Kingdom, Ireland, Greece, Denmark, Portugal, and later Sweden and Finland) on the grounds that they did not have sub-national tiers of government above local or municipal level (Loughlin, 1997, p. 157). The principal point at issue has been the status in their home countries of the members nominated by national governments. It was the clearly expressed wish of the Commission, and subsequently of the European Parliament, that CoR members should be elected representatives within their own regions, so that they could be seen to have legitimate democratic credentials (*JOCE*, noC58, 1.3.1993, p. 26). The Council of Ministers, on the other hand, merely advised that states should choose members according to their suitability to represent local and regional authorities. The initial response from the United Kingdom and Greece, for example, was to attempt to nominate unelected civil servants, although subsequent shifts in policy led to the appointment of a variety of (elected) representatives of local authorities. An interesting insight into the functioning of European institutions can be gained from the fact that the pretext for bypassing elected representatives in the first instance in the United Kingdom and Greece arose from tendentious English and Greek language versions of the Maastricht Treaty. References in English to 'representatives of regional and local bodies' and, in Greek, to 'regional administrations' provided the loophole to justify the appointment of unelected members. Faced with these anomalies, the Commission rapidly closed the loophole to bring the two aberrant versions into line with those in other languages (Féral, 1998, p. 28). A large majority of nominated CoR members are now therefore elected within their own regional or local constituencies. They are, none the less, expected to devote their energies while in office to the interests of the Committee and the European Union as a whole, and not to narrow regional or sectoral interests. It is therefore salutary to note the widely-held view that 'the primary divisions within the CoR have so far been overwhelmingly along national lines' (Collins & Jeffery, 1997, p. 15).

In territorial terms, the origins of CoR members are also subject to immense variation according to country. In the simplest case, Luxembourg, with no regional tier of government, a single level of representation at local authority level suffices. Most other states have opted for a mixture of regional, local/municipal and intermediate representation with, for example, a communal, departmental and regional distribution in France and a complex pattern for the United Kingdom with representation following differing patterns in England, Scotland, Wales and Northern Ireland.

In view of this complex pattern of nomination and representation, it was to be expected that decisions about the workings of the Committee and in particular its presidency, would prove to be sensitive. For its first four-year term (1994-8), agreement was reached that led to the election of Jacques Blanc, President of France's Languedoc-Roussillon region, with Pasquall i Maragall, Mayor of Barcelona, as his deputy, on the understanding that the positions of president and deputy would be reversed after two years. Maragall was thus elected to the presidency in March 1996. This arrangement served the purpose of guaranteeing equal influence at a high level to representatives of both regional and municipal/local authorities. It also produced a political balance between the centre-right (Blanc) and the left (Maragall). With the renewal of the Committee's mandate for a second four-year term from March 1998, a similar pattern has emerged. The German Social Democrat, Manfred Dammeyer, Minister for Federal and European Affairs in North Rhine Westphalia, has been elected President until 2000, when his deputy, Josef Chabert, centre-right Finance Minister for Brussels Capitale region, will assume the presidency for the remaining two years.

From the inception of the Committee of the Regions, a certain number of policy areas of direct regional interest were designated as those on which it is obligatorily consulted by the decision-making bodies of the European Union – education; culture; public health; trans-European networks for transport, telecommunications and energy; economic and social cohesion. It is entitled to offer opinions on any other subject that it deems relevant to regional concerns. There is, however, no obligation on any other European institution to take account of the opinions expressed. In view of its relatively limited sphere of influence, both in terms of the range of subjects on which it was entitled to produce opinions and of the weight attached to those opinions, the CoR has been predictably keen to increase its range and to gain greater influence. The series of consultations and negotiations which took place under the heading of the Inter-Governmental Conference (IGC), culminating in the Treaty of Amsterdam signed in June 1997, provided the opportunity for intensive lobbying on the part of the CoR. The Catalan Prime Minister, Jordi Pujol, oversaw the preparation of a report subsequently submitted as an official Opinion for the IGC and which sought to defend and reinforce the Committee's status among the Community institutions. Among the measures proposed were an extension of those areas in which the CoR is obligatorily consulted on matters of policy, to include among others aspects of agriculture, social policy, transport, vocational training, environmental protection. Perhaps most significant was a proposal put forward for the benefit of regional and local tiers of government rather than in the interests of the CoR itself. This was an attempt to reformulate the principle of subsidiarity so that it would extend automatically to the sub-national level of responsibility, 'the Community shall take action, in accordance with the principle of subsidiarity, only if and so far as the objectives of the proposed action cannot be sufficiently achieved by the Member States, or by the regional and local authorities endowed with powers under the domestic legislation of the Member State in question' (CdR 136/95, 20-21 April 1995, p.9). The final element of the lobbying process that preceded the culmination of the IGC was a 'summit' conference organised by the CoR in Amsterdam one month before the European Council meeting which was to complete the revision of the Community Treaties in that city in June 1997. This conference, which had as its basic text a report, 'Regions and Cities, pillars of Europe' prepared by the Prime Minister of Bavaria and the Mayor of Oporto, was designed to underline the importance of local and regional authorities.

In the event, the Amsterdam Treaty of 1997 produced modest gains for the Committee of the Regions. The range of subjects on which it must be consulted has been enlarged to include aspects of employment, social policy, health, the environment, vocational training and transport. It has also been granted the status of 'expert' on matters concerning cross-border cooperation. In addition, it has gained a greater measure of administrative freedom to the extent that is now permitted to develop its own internal regulations. It may be claimed, therefore, that the Committee of the Regions is now more firmly embedded among the institutional structures of the European Union. Prospects for the years to come suggest that it will need to be. Already negotiations are underway which are destined to lead to enlargement of the European Union. There is no shortage of supplicants among the countries of Central and Eastern Europe and the Mediterranean zone: the Czech Republic, Estonia, Hungary, Poland, Slovenia, Bulgaria, Latvia, Lithuania, Slovakia, Romania, Cyprus, Malta, although progress towards enlargement is not likely to be rapid, particularly in view of the impact on the Union's budget of the accession of countries which are economically backward compared with the European Union average. The regional dimension, however, is unlikely to diminish in importance. The citizens of Europe, whatever its size, all belong somewhere, and a subtle and sensitive balance of institutions and communities is required to satisfy their sense of belonging and to accommodate their aspirations and needs. As events over the last decade have demonstrated, states, nations and their institutions need to adapt rapidly to survive, to flourish, and to cope with the endlessly complex and evolving patterns of allegiance and identity.

References

Bourrinet, J. (ed.) (1997), *Le Comité des regions de l'Union européenne*, Paris, Economica.

Collins, S., & C. Jeffery (1997), *Whither the Committee of the Regions? British and German Perspectives*, London, Anglo-German Foundation.

Feral, P.-A. (1998), *Le Comité des régions de l'Union européenne*, Paris, PUF.

Loughlin, J. (1997), 'Representing Regions in Europe: The Committee of the Regions', in Jeffery, C. (ed.), *The Regional Dimension of the European Union: Towards a Third Level in Europe*, London & Portland, Or., Frank Cass, pp. 147-165.

Bibliography

Adamson, D. L. (1991), *Class, Ideology and the Nation*, Cardiff, University of Wales Press.

Allum, P. A. (1973), *Italy: Republic without Government?* New York, Norton.

Alonso Fernández, J. (1993), *La Nueva Situación Regional*, Madrid, Editorial Síntesis.

Andrén, N. (1967), 'Nordic Integration', *Cooperation and Conflict*, 2 (1), 1-25.

Andrikopoulou, E. (1992), 'Whither Regional Policy? Local Development and the State in Greece' in Dunford, M. & G. Kafkalas (eds), *Cities and Regions in the New Europe: The Global-Local Interplay and Spatial Development Strategies*, London, Belhaven Press.

Archer, C. (1998), 'The Baltic-Nordic Region' in Park, W. & G. Wyn Rees (eds), *Rethinking Security in Post-Cold War Europe*, Longman, Harlow.

Aron, R. (1977), *Léopold III ou le choix impossible: février 1934 - juillet 1940*, Paris, Plon.

Ascherson, N. (1988), *Games with Shadows*, London, Radius.

Audit Commission (1991), *A Rough Guide to Europe: Local Authorities and the EC*, London, HMSO.

Bachtler, J. & C. Clement (1992), 'Community regional policies for the 1990s', in *Regional Studies*, 26 (3), 414.

Bagnasco, A. (1977), *Tre Italie. La problematica territoriale dello sviluppo italiano*, Bologna, Il Mulino.

Barrington, T. (1976), 'Can there be Regional Development in Ireland?', *Administration*, 24, 350-65.

— (1982), 'Whatever Happened to Irish Government?', in Litton, F. (ed.), *Unequal Achievement: The Irish Experience 1957-1982*, Dublin, Institute of Public Administration, pp. 89-112.

Battye, A. & M.-A. Hintze (1992), *The French Language Today*, London, Routledge.

Becchi, A. (1992), 'Incentivi statali, pesante eredità', in *A sud di qualunque nord*, Il Manifesto del mese, n. 8.

Benko, G., & A. Lipietz (eds) (1992), *Les Régions qui gagnent: districts et réseaux: les nouveaux paradigmes de la géographie économique*, Paris, PUF.

Benz, A. (1993), 'Redrawing the Map? The Question of Territorial Reform in the Federal Republic', in Jeffery, C. & R. Sturm (eds), *Federalism, Unification and European Integration*, London, Frank Cass, 38-57.

Bernecker, W. L. (1990), 'Spain and Portugal between regime transition and stabilized democracy', *Iberian Studies*, 19, 32-56.

Bogdanor, V. (1979), *Devolution*, Harmondsworth, Penguin.

Borkenhagen, F., *et al* (eds) (1992), *Die Deutschen Länder in Europa*, Baden-Baden, Nomos.

Bosch Gimpera, P. (1996), *El problema de las Españas*, Málaga, Algaraza.

Bossi, U. (with D. Vimercati) (1992), *Vento dal Nord. La mia Lega la mia vita*, Milan, Sperling & Kupfer.

Bourrinet, J. (ed.) (1997), *Le Comité des regions de l'Union européenne*, Paris, Economica.

Brand, J. (1978), *The National Movement in Scotland*, London, Routledge & Kegan Paul.

Braudel, F. (1986), *L'Identité de la France: espace et histoire*, Paris, Arthaud-Flammarion.

Brongniart, P. (1971), *La Région en France*, Paris, Colin.

Bronner, O. (ed.), '*Der Standard*' – Österreichs unabhängige Tageszeitung für Wirtschaft, Politik und Kultur, 7 January 1997.

Brunt, B. (1988), *The Republic of Ireland*, London, Paul Chapman.

— (1993), 'Ireland as a Peripheral Region of Europe: Structural Funds and Regional Economic Development', in King, R. (ed.), *Ireland, Europe and the Single Market*, Dublin, Geographical Society of Ireland, pp. 30-43.

Buchanan Report (1968), *Regional Studies in Ireland*, Dublin, Stationery Office.

Budd, L. (1997), 'Regional Government and Performance in France', *Regional Studies*, 31, 2 (April), 187-192.

Bundesrat (1989), *Vierzig Jahre Bundesrat*, Baden-Baden, Nomos.

Bundesrat (1991), *Handbuch des Bundesrates*, Munich, Beck.

Burke, E. (1905), *Reflections on the French Revolution*, London, Methuen.

Butler, D. & D. Kavanagh (1988), *The British General Election of 1987*, London, Macmillan.

— (1992), *The British General Election of 1992*, London, Macmillan.

Butt Philip, A. (1975), *The Welsh Question*, Cardiff, University of Wales Press.

— (1978), *Creating New Jobs*, London, Policy Studies Institute.

Cammelli, M. 'Regioni e rappresentanza degli interessi: il caso italiano', *Stato e mercato* , N. 2. August 1990, 151-200.

Capon, B, *The Times*, 2 November 1993.

Childs, M. W. (1980), *The Middle Way on Trial*, New Haven, Yale University Press.

Christakis, M. (1998), 'Greece: Competing Regional Priorities' in Hanf, K. & B. Soetendorp (eds), *Adapting to European Integration: Small States and the European Union,* London, Longman, pp. 84-99.

Christopoulos, D. (1998), 'Clientelistic Networks and Local Corruption: Evidence from Western Crete', *South European Politics and Society*, 3 (1), 1-22.

Chubb, J. (1990), *Patronage, Power and Poverty in Southern Italy,* Cambridge, Cambridge University Press.

Clark, M. (1983), *Modern Italy 1871-1982*, London & New York, Longman.

Clark, R. P. (1987), 'The question of regional autonomy in Spain's democratic transition', in Clark, R.P. & M. H. Haltzel (eds), *Spain in the 1980s*, Cambridge, Mass., Ballinger.

Clement, C. (1991), 'Regional economic policy for the next few years', *Tijdschrift vor Econ. En Soc. Geografie*, 82, (3), 228.

Clogg, R. (1992), *A Concise History of Modern Greece*, Cambridge, Cambridge University Press.

Club Jean Moulin (1968), *Les Citoyens au pouvoir: 12 régions, 2000 communes*, Paris, Le Seuil.

Cole, J. & F. Cole (1993), *The Geography of the European Community*, London & New York, Routledge.

Collins, S. & C. Jeffery (1997), *Whither the Committee of the Regions? British and German Perspectives*, London, Anglo-German Foundation.

Conversi, D. (1997), *The Basques, the Catalans and Spain: alternative routes to nationalist mobilisation*, London, Hurst.

Corkhill, D. (1992), 'Imperfect bipolarism? Portugal's political system after the 1991 Parliamentary election', *Journal of the Association for Contemporary Iberian Studies*, 5, 16-23.

Coutinho, A. B. (1978), *Que Futuro para os Açores?* Lisbon, Editorial Caminho.

CSS (1991), *Italy today: social picture and trends, 1990*, Milan, Angeli.

Culliton Report (1992), *A Time for Change: Industrial Policy for the 1990s. Report for the Industrial Policy Review Group*, Dublin, Stationery Office.

Currie, D. P. (1994), *The Constitution of the Federal Republic of Germany*, Chicago & London, University of Chicago Press.

Dayries, J.-J. & M. Dayries (1978), *La Régionalisation*, Paris, PUF.

De Mauro, T. (1963), *Storia linguistica dell'Italia unita*, Bari, Laterza.

Delpino, L. & F. Del Giudice (1992), *Diritto amministrativo*, Naples, Simone.

Dematteis, G. (1989), 'Regioni geografiche, articolazione territoriale degli interessi e regioni istituzionali', *Stato e mercato*, 27.

Demertzis, N. (1997), 'Greece', in Eatwell, R. (ed.), *European Political Cultures: Conflict or Convergence,* London, Routledge, pp.107-121.

Department of Employment, *Preliminary Results from the 1987 Labour Force Survey* (1988), Table 4.

Department of Trade and Industry, *DTI – The Department of Enterprise*, London HMSO, Cmnd. 278, January 1988.

Diamandouros, N. (1983), 'Greek Political Culture in Transition: Historical Origins, Evolution Current Trends' in Clogg, R. (ed.), *Greece in the 1980s*, New York, St. Martin's Press, pp. 43-69.

Dijkink, G. (1995), 'Metropolitan government as a political pet? Realism and tradition in administrative reform in the Netherlands', *Political Geography*, 14 (4), 335.

Donaghy, P. J. & M. T. Newton (1987), *Spain: A Guide To Political and Economic Institutions*, Cambridge, Cambridge University Press.

Donzelli, C. (1992), 'Le oscillazioni del federalismo', in *A sud di qualunque nord*, Il Manifesto del mese, n. 8.

Downs, W. M. (1996), 'Federalism achieved: The Belgian Elections of May 1995', *West European Politics*, 19 (1), 168-175.

Drevet, J.-F. (1988), *1992-2000: les régions françaises entre l'Europe et le déclin*, Paris, Souffles.

— (1991), *La France et l'Europe des régions*, Paris, Syros.

Dumont, G.-H. (1991), *La Belgique*, Paris, PUF.

Earle, J. (1974), *Italy in the 1970s*, Newton Abbot & Vancouver, David & Charles.

Economic and Social Committee of the European Communities (1983), *Irish Border Areas: Information Report*, Brussels, General Secretariat of the Economic and Social Committee.

The Economist, 26 June 1993.

The Economist, 20 September 1997.

EIU–The Economist Intelligence Unit (1998), 'Quarterly Report: April 24, 1998', *Country Report: Greece 1998-99*, London, EIU.

ESRI–Economic and Social Research Institute (1997), *Aggregate and Regional Impact: The Cases of Greece, Spain, Ireland and Portugal*, Luxembourg, Office for Official Publications of the European Communities.

EU Consequences Commission on Municipalities and County Councils (1994), *The Municipalities, the County Councils and Europe*, Stockholm, Norstedts.

Exler, U. (1993), 'Financing German Federalism: Problems of Financial Equalisation in the Unification Process', in Jeffery, C. & R. Sturm (eds), *Federalism, Unification and European Integration*, London, Frank Cass, pp. 22-37.

European Commission (1995), *Sweden: Objective 6 Single Programming Document 1995-1999*, Directorate General for Regional Policies and Cohesion No. 95.15.13.001, European Commission, Brussels.

Faludi, A. & A.Van der Valk (1994), 'Provincial planning the lynchpin?' in *Rule and Order: Dutch Planning Doctrine in the Twentieth Century*, London, Kluver.

Falzone, V., F. Palermo & F. Cosentino (eds) (1976), *La Costituzione della Repubblica Italiana. Illustrata con i lavori preparatori*, Milan, Mondadori.

Featherstone, K. & C. Ifantis (eds) (1996), *Greece in a Changing Europe: Between European Integration and Balkan Disintegration?* Manchester, Manchester University Press.

Feijo, R. C. (1989), 'State, nation and regional diversity in Portugal: an overview', in Herr, R. & J.H.R. Polt (eds), *Iberian Identity: Essays on the Nature of Identity in Portugal and Spain*, Berkeley, Institute of International Studies.

Fennell, D. (1983), *The State of the Nation: Ireland since the Sixties*, Dublin, Ward River.

Feral, P.-A. (1998), *Le Comité des régions de l'Union européenne, Paris, PUF.*

Feron, F. & A. Thoraval (eds) (1992), *L'Etat de l'Europe*, Paris, La Découverte.

Fitzmaurice, J. (1984), 'Belgium: Reluctant Federalism', *Parliamentary Affairs*, 37, 418-433.

— (1996), *The Politics of Belgium: A Unique Federalism*, London, Hurst.

Fourastié, J. (1979), *Les trente glorieuses: ou la révolution invisible de 1946 à 1975*, Paris, Fayard.

Frenkel, M, (1984/86), *Föderalismus und Bundesstaat* (2 vols), Bern/Frankfurt.

Frognier, A. P., M. Quevit & M. Stenbock (1982), 'Regional Imbalances and Centre-Periphery

Relations in Belgium', in Rokkan, S. & D. W. Urwin (eds), *The Politics of Territorial Identity: Studies in European Regionalism.*

Fuà, G. & C. Zacchia (eds), *Industrializzazione senza fratture*, Bologna, Il Mulino.

Gallagher, T. (1976), 'Portugal's Atlantic Territories: The Separatist Challenge', *The World Today*, pp. 353-360.

— (1979), 'Portugal's relations with her Atlantic territories', *The World Today*, March.

García Ferrando, M. (1982), *Regionalismo y autonomías en España, 1976-1979*, Madrid, Centro de Investigaciones Sociológicas.

García Ferrando, M., E. López Aranguren & M. Beltrán (1994), *La conciencia nacionalista y regional en la España de las autonomías*, Madrid, Centro de Investigaciones Sociológicas.

Gaspar, J. (1984), 'The unity and individuality of Portugal', *Iberian Studies*, 13, 3-7.

— (1990), 'The new map of Portugal', in Hebbert, M. & J. C. Hansen (eds), *Unfamiliar Territory: The Reshaping of European Geography*, Aldershot, Avebury Press.

Gilbert, M. (1998), 'Transforming Italy's Institutions? The Bicameral Committee on Institutional Reform', *Modern Italy,* 3 (1), 49-66.

Gilmore, D. D. (1981), 'Andalusian regionalism: anthropological perspectives', *Iberian Studies*, 10, 58-67.

Ginsborg, P. (1990), *A History of Contemporary Italy: Society and Politics*, London, Penguin.

Gravier, J.-F. (1947), *Paris et le désert français*, Paris, Portulan.

— (1964), *L'Aménagement du territoire et l'avenir des régions françaises*, Paris, Flammarion.

— (1970), *La Question régionale*, Paris, Flammarion.

Greenwood, J.(1997), *Representing interests in the European Union*, New York, St Martin's Press.

Gruber, A. (1986), *La Décentralisation et les institutions administratives*, Paris, Colin.

Guerra, L. L. (1989), 'National and regional pluralism in Contemporary Spain', in Herr, R. & J.H.R. Polt (eds), *Iberian Identity: Essays on the Nature of Identity in Portugal and Spain*, Berkeley, Institute of International Studies.

Gunlicks, A. B. (1995), 'The "Old" and the "New" Federalism', in Merkl, P. H. (ed.), *The Federal Republic of Germany at Forty-Five*, London, Macmillan, pp. 219-242.

Häggroth, S., K. Kronvall, C. Riberdahl & K. Rudebeck (1996), *Swedish Local Government: Traditions and Reforms,* Stockholm, Svenska Institutet.

Haller, M. (1996), *Identität und Nationalstolz der Österreicher*, Vienna, Böhlau.

Hamann, B. (1996), *Hitlers Wien. Lehrjahre eines Diktators*, Munich, Piper, 1996.

Hanham, H. (1969), *Scottish Nationalism*, London, Faber.

Hanisch, E. (1992), 'Kontinuitäten und Brüche: Die innere Geschichte', in Dachs, H. *et al* (eds), *Handbuch des politischen Systems Österreichs*, Vienna, Manz.

Hardy, S., M. Hart *et alii* (eds) (1995), *An Enlarged Europe. Regions in Competition?* London, Regional Studies Association.

Harvey, D. (1989), 'From managerialism to entrepreneurialism: the transformation of urban governance in late capitalism', *Geografiska Annaler*, 71b, 3-17.

Hasquin, H. (1982), *Historiographie et politique: Essai sur l'histoire de la Belgique et la Wallonie*, 2nd edition, Charleroi, Institut Jules Destrée.

Haycraft, J. (1985), *Italian Labyrinth*, Harmondsworth, Penguin.

Hebbert, M. (1990), 'Spain – A Centre Periphery Transformation', in Hebbert, M. & J.C. Hansen (eds), *The Reshaping of European Geography*, Aldershot, Avebury Press.

Hencke, D. 'Welfare cuts hit poorest areas worst', *The Guardian*, 20 September 1993.

Hendriks, F. (1997), 'Regional reform in the Netherlands: reorganising the viscous state', in Keating, M. & J. Loughlin (eds), *The Political Economy of Regionalism*, pp. 370-387.

Henri, comte de Paris (1996), *Les Rois de France et le sacré*, Monaco, Editions du Rocher.

Herzfeld, M. (1985), *The Poetics of Manhood: Contest and Identity in a Cretan Mountain Village*, Princeton, New Jersey, Princeton University Press.

Hesse, K. (1984), *Grundzüge des Verfassungsrechts des Bundesrepublik Deutschland*, Karlsruhe, Müller.

Hine, D. (1993), *Governing Italy. The Politics of Bargained Pluralism*, Oxford, Clarendon Press.

House of Lords Select Committee on the European Communities, *European Regional Development Fund*, 23rd Report of Session 1983-84, HL Paper 274, pp. xxix-xxxii.

House of Lords Select Committee on the European Communities, *Reform of the Structural Funds*, 14th Session 1987-88, HL Paper 82, pp. 18-19.

Hrbek, R. & U. Thaysen (eds) (1986), *Die Deutschen Länder und die Europäischen Gemeinschaften*, Baden-Baden, Nomos.

Huggett, F. E. (1969), *Modern Belgium*, London, Pall Mall Press.

Hutton, W. (1995), *The State we're in*, London, Cape.

Ioakimidis, P.C. (1996), 'EU Cohesion Policy in Greece: The Tension Between Bureaucratic Centralism and Regionalism', in Hooghe, L. (ed.), *Cohesion Policy and European Integration: Building Multi-Level Governance*, Oxford, Oxford University Press, pp. 342-66.

Ignatieff, M. (1993), *Blood and Belonging: Journeys into the New Nationalism*, London, BBC & Chatto & Windus.

Italia, V. & M. Bassani (eds) (1990), *Le autonomie locali. Legge 8 giugno 1990, n. 142)*, Milan, Giuffrè.

Jeffery, C. (ed.) (1997), *The Regional Dimension of the European Union: Towards a Third Level in Europe?*, London, Portland, Or., Frank Cass.

Jeffery, C. (1998), 'German Federalism in the 1990s: On the Road to a "Divided Polity"?', in Larres, K. (ed.), *Germany since Unification*, London, Macmillan, pp. 107-128.

Jones, B. & M. Keating (eds) (1995), *The European Union and the Regions*, Oxford, Clarendon Press.

Kazakos, P. & P. C. Ioakimidis (eds) (1994), *Greece and EC Membership Evaluated*, London, Pinter.

Keating, M. (1988), *State and Regional Nationalism: Territorial Politics and the European State*, London, Harvester Wheatsheaf.

— (1992), 'Regional Autonomy in the Changing State Order: A Framework of Analysis', *Regional Politics and Policy*, 2 (3).

Keating, M. & P. Hainsworth (1986), *Decentralization and Change in Contemporary France*, Aldershot, Gower.

Keating, M. & J. Loughlin (1997), *The Political Economy of Regionalism*, London, Frank Cass.

Kedourie, E. (1960), *Nationalism*, London, Hutchinson.

Kennedy, K., T. Giblin & D. McHugh (1988), *The Economic Development of Ireland in the Twentieth Century*, London & New York, Routledge.

Kilper, H. & R. Lhotta (1996), *Föderalismus in der Bundesrepublik Deutschland: eine Einführung*, Opladen, Leske & Budrich.

Kockel, U. (1990), *Dublin's Inner City: Community-based Initiatives and Employment*, Occasional Papers in Irish Studies 1, University of Liverpool (with M. Ross).

— (1993), *The Gentle Subversion: Informal Economy and Regional Development in the West of Ireland*, Bremen, European Society for Irish Studies.

Konstadakopulos D. & D. Christopoulos (1997), 'Innovative Milieux and Networks, Technological Change and Learning in European Regions', presented at the international conference 'Technology Policy and Research on Development Systems in Europe', United Nations University, Institute for New Technologies, Seville, October 1997, mimeo., 28 pp.

Kossmann-Putto, J. A. & E. H. Kossmann (1987), *The Low Countries: History of the Northern and Southern Netherlands*, Flanders, Flemish Netherlands Foundation.

Kostopoulos, T. (1996), *Europaiki Enosi kai Topiki Autodiikisi. Tomos Protos: O Thesmos tis Topikis Autodiikisis stin Ellada* [*European Union and Local Government. Volume One: The Institution of Local Government in Greece*], Athens, Ekdosis Papazisi.

Lafont, R. (1967), *La Révolution régionaliste*, Paris, Gallimard.

— (1971), *Décoloniser en France*, Paris, Gallimard.

Lanversin, J. de, A. Lanza, F. Zitouni (1989), *La Région et l'Aménagement du territoire dans la décentralisation*, 4th edition, Paris, Economica.

Larsson, T. (1995), *Governing Sweden,* Stockholm, Statskontoret.

Laufer, H. & F. Pilz (1973), *Föderalismus*, Munich.

Lausen, J. R. (1986), *El Estado Multi-Regional – España Descentrada*, Madrid, Alianza Editorial.

Le Bras, H. & E. Todd (1981), *L'Invention de la France: atlas anthropologique et politique*, Paris, Librairie Générale Française.

Lega Lombarda-Lega Nord, *Programma elettorale* , February 1992.

Lehmbruch, G. (1976), *Parteienwettbewerb im Bundesstaat*, Stuttgart, Kohlhammer.

Leonardi, R., R. Nanetti & R. Putnam (1987), 'Italy: Territorial Politics in the Postwar Years: The Case of Regional Reform', *West European Politics*, 10 (4), 88-107.

Lepschy, A. L. & G. Lepschy (1991), *The Italian Language Today*, London, Routledge.

Lepschy, A. L., G. Lepschy & M. Voghera (1993), 'Linguistic Variety in Italy', paper given at the 1992 Annual Conference of the Association for the Study of Modern Italy [ASMI]. A summary of this paper appeared in the ASMI Newsletter, N. 23, Spring 1993.

Lévy, M. L. (1982), *La Population de la France des années 80*, Paris, Hatier.

Lewis, J. R. & A. M. Williams (1981), 'Regional uneven development on the European periphery: the case of Portugal', *Tijdschirft voor Economische en Sociale Geografie*, 72, 81-98.

— (1984), 'Social cleavages and electoral performance: the social basis of Portuguese political parties, 1976-1983', *West European Politics*, 7 (2), 119-137.

— (1994), 'Regional autonomy and the European Communities: the view from Portugal's Atlantic islands', *Regional Politics and Policy*, 4 (2), 67-85.

Logie, J. (1980), *1830: De la régionalisation à l'indépendance*, Paris, Duculot.

López, C.E.D. (1982), 'The politicization of Galician cleavages', in Rokkan, S. & D. W.Urwin (eds), *The Politics of Territorial Identity: Studies in European Regionalism*, London, Sage.

López Aranguren, E. (1994), 'Nacionalismo, Regionalismo y Postnacionalismo en las Comunidades Autónomas del Estado Español', *Razón y Fe*, 230 (1153), 269-281.

Loughlin, J. (1997), 'Representing Regions in Europe: The Committee of the Regions', in Jeffery, C. (ed.), *The Regional Dimension of the European Union: Towards a Third Level in Europe*, London & Portland, Or., Frank Cass, pp. 147-165.

— (1998), 'The Regional Question In Europe: An Overview', Paper presented to the 'Federalism and European Union' Conference, Centre for EU Studies, University of Hull (March 1998).

Loughlin, J. & S. Mazey (eds) (1995), *The End of the French Unitary State? Ten Years of Regionalization in France (1982-1992),* London, Frank Cass.

Lynch, P. (1996), *Minority Nationalism and European Integration*, Cardiff, University of Wales Press.

Lyrintzis, C. (1984), 'Political Parties in Post Junta Greece: A Case of "Bureaucratic Clientelism"?' *West European Politics*, 7 (2), 98-118.

Mabille, X. (1986), *Histoire politique de la Belgique: facteurs et acteurs de changement*, Brussels, Centre de recherche et d'information socio-politiques.

McAlinden, G. (1995), 'The European Union: A Better Life on the Border', in D'Arcy, M. & T. Dickson (eds), *Border Crossings: Developing Ireland's Island Economy*, Dublin, Gill & Macmillan, pp. 77-84.

Mack Smith, D. (1985), *Cavour*, London, Methuen.

— (ed.) (1968), *The Making of Italy*, New York, Harper & Row.

Maldonado Gago, J. (1995), 'España, una nación de naciones', *Política y Sociedad*, 20, 23-33.

Mannheimer, R. (ed.) (1991), *La Lega lombarda*, Milan, Feltrinelli.

Mar-Molinero, C., & A. Smith (eds) (1996), *Nationalism and the nation in the Iberian peninsula: competing and conflicting identities*, Oxford, Berg.

Marko, J. (1992), 'Die Verfassungssysteme der Bundesländer: Institutionen und Verfahren repräsentativer und direkter Demokratie', in Dachs, H. *et al* (eds), *Handbuch des politischen Systems Österreichs*, Vienna, Manz.

Markovits, A.S. (1996), 'Austrian corporatism in comparative perspective', in Bischof, G. & A. Pelinka (eds), *Austro-Corporatism, past, present, future*, (Contemporary Austrian Studies, vol. 4), New Brunswick & London, Transaction Publishers, pp. 5-20.

Marr, A. (1996), *Ruling Britannia: the failure and future of British democracy*, London, Penguin.

Mazower, M. (1998), *Dark Continent: Europe's Twentieth Century*, London, Allen Lane The Penguin Press.

Meijer, H. (1989), *The Fourth Policy Document on Physical Planning in the Netherlands*, Utrecht, Den Haag, IDC.

Mény, Y. (1974), *Centralisation et décentralisation dans le débat politique français (1945-1969)*, Paris, Pichon & Durand-Auzias.

Merriman, J. M. (ed.) (1982), *French Cities in the Nineteenth Century*, London, Hutchinson.

Miles, L. (1997a), *Sweden and European Integration*, Ashgate, Aldershot.

— (1997b), 'Sweden: A Relevant or Redundant Parliament', *Parliamentary Affairs*, 50 (3), 423-37.

Moores, B., J. Rhodes & P. Tyler (1986), *The Effects of Government Regional Policy*, London, Department of Trade and Industry, HMSO.

Morata, F. (1995), 'Spanish Regions in the European Community', in Jones, B. & M. Keating (eds), *The European Union and the Regions*, pp. 114-133.

— (1997), 'The Euro-region and the C-6 Network: The New Politics of Sub-national Cooperation in the West-Mediterranean Area', in Keating, M. & J. Loughlin (eds), *The Political Economy of Regionalism*, pp. 292-305.

Morisi, M. (ed.) (1987), *Regioni e rappresentanza politica*, Milan, Angeli.

Mughan, A. (1983), 'Accommodation or Diffusion in the Management of Linguistic Conflict in Belgium', *Political Studies*, 1983, XXXI, 434-451.

Munck, R. (1993), *The Irish Economy: Results and Prospects*, London, Pluto.

Murphy, A. (1995), 'Belgium's Regional Divergence: Along the Road to Federation', in Smith, G., *Federalism: The Multiethnic Challenge*, London, Longman, pp. 73-100.

Nairn, T. (1977), *The Break-up of Britain: Crisis and Neo-Nationalism*, London, NLB.

Nanetti, R. (1988), *Growth and Territorial Policies. The Italian Model of Social Capitalism*, London & New York, Pinter.

Newhouse, J. (1997), 'Europe's Rising Regionalism', *Foreign Affairs*, 76, 1 (January / February), 67-84.

Newton, M. (1982), 'Andalusia: The Long Road to Autonomy', *Journal of Area Studies*, 6, 27-32.

— (1983), 'The peoples and regions of Spain', in Bell, D. S. (ed.), *Democratic Politics in Spain*, London, Pinter.

Nick, R. & A. Penlinka (1992), *Österreichs politische Landschaft*, Innsbruck, Haymon.

Nyman, O. (1960), *Der westdeutsche Föderalismus*, Stockholm.

Osmond, J. (ed.) (1985), *The National Question Again*, Llandysul, Gomer Press.

Ossenbühl, F. (ed.) (1990), *Föderalismus und Regionalismus in Europa*, Baden-Baden, Nomos.

Ó Tuathaigh, G. (1986), 'The Regional Dimension', in Kennedy, K. (ed.) *Ireland in Transition: economic and social change since 1960*, Cork & Dublin, Mercier, pp.120-32.

Pagenkopf, H. (1981), *Der Finanzausgleich im Bundesstaat*, Stuttgart, Kohlhammer.

Payne, S. (1971), 'Catalan and Basque nationalism', *Journal of Contemporary History*, 6, 15-51.

Pereira, A. (1995), 'Regionalism in Portugal', in Jones, B. & M. Keating (eds), *The European Union and the Regions*, pp. 268-280.

Perrineau, P. (1987), *Régions: le baptême des urnes*, Paris, Pedone.

Petersson, O. (1994), *Swedish Government and Politics*, Stockholm, Publica.

Pi-Sunyer, O. (1988), 'Catalan politics and Spanish democracy: an overview of a relationship', *Iberian Studies*, 17, 1-16.

Pinha-Cabral, J. de (1989), 'Sociocultural differentiation and regional identity in Portugal', in Herr, R. & J.H.R. Polt (eds), *Iberian Identity: Essays on the Nature of Identity in Portugal and Spain*, Berkeley, Institute of International Studies.

Podbielski, G. (1974), *Italy: Development and Crisis in the Post-war Economy*, Oxford, Clarendon Press.

Poole, A. (1987), 'The Fourons: a microcosm of Belgium's linguistic problems', *The Linguist*, 26 (2), 52-56.

Porto, M. (1984), 'Regional development in Portugal: the institutional framework', *Iberian Studies*, 13, 7-16.

Putnam, R. D. (with R. Leonardi & R. Nanetti) (1993), *Making Democracy Work. Civic Traditions in Modern Italy*, Princeton, New Jersey, Princeton University Press.

Putnam, R.D., R. Leonardi & R. Nanetti (1985), *La pianta e le radici*, Bologna, Il Mulino.

Quartermaine, L. & J. Pollard (eds) (1985), *Italy Today: Patterns of Life and Politics*, Exeter, University of Exeter.

Ragionieri, E. (1976), *La storia politica e sociale*, Storia d'Italia, Vol 4:3, Turin, Einaudi.

Ragionieri, E. (1979), *Politica e amministrazione nella storia dell'Italia unita*, Rome, Editori Riuniti.

Regeringskansliet (1988), *Kommuner Kan! Kanske! – om kommunal välfärd I framtiden,* Ds 1998, 15, Stockholm, Regeringskansliet.

Report of the Royal Commission on Local Government in England (1969), Cmnd. 4040, London, HMSO.

Report of the Royal Commission on the Constitution (1973), Vols I and II, Cmnd. 5460, London, HMSO.

La Repubblica, 1 April 1992.

Rhodes, R.A.W. & V. Wright (1987), 'Introduction', *West European Politics*, 10 (4) (Special Issue on 'Tensions in the Territorial Politics of Western Europe'), 1-20.

Ritaine, E. (1998), 'The Political Capacity of Southern European Regions', in Le Gales, P. & C. Lequesne (eds), *Regions in Europe*, London, Routledge, pp. 76-88.

Rodotà, C. (1986), *La Corte Costituzionale*, Rome, Editori Riuniti.

Rokkan, S. & D. W. Urwin (eds) (1982), *The Politics of Territorial Identity: Studies in European Regionalism*, London, Sage.

Rose, R. (1982), *Understanding the United Kingdom: The Territorial Dimension*, Harlow, Longman.

Ross, C. (1991), 'Towards the Basque election of 1990: the nationalist realignment of 1986-1989 and its effects', *Journal of the Association of Contemporary Iberian Studies*, 4, 49-59.

Rubiralta Casas, F. (1997), 'Le nouveau nationalisme radical dans l'Etat espagnol: analyse comparative d'un processus univoque en Galicie, en Catalogne et au Pays Basque', *Revue International de Politique Comparée*, 4 (2), 459-472.

Ruffilli, R. (1971), *La questione regionale dall'unificazione alla dittatura* , Milan, Giuffrè.

Rugge, F. (1997), 'Le leggi "Bassanini": continuità e innovazioni del riformismo amministrativo', *il Mulino*, 4 (July-August), 717-26.

Rupeni, A. (ed.) (1980), *Comuni e Province negli anni 80*, Rome, Cinque Lune.

Sampson, A. (1992), *The Essential Anatomy of Britain: Democracy in Crisis*, London, Hodder & Stoughton.

Sassoon, D. (1986), *Contemporary Italy*, London, Longman.

Scheuch, M. (1995), *Österreich, Provinz Weltreich, Republik*, Vienna, Brandstätter.

Schmidt, V. A. (1990), *Democratizing France: the political and administrative history of decentralization*, Cambridge, Cambridge University Press.

Senelle, R. (1987), *The Reform of the Belgian State*, Vol. IV, Brussels, Ministry of Foreign Affairs and External Trade (Memo from Belgium No. 196).

Sieder, R., H. Steinert & E. Tálos (eds) (1995), *Österreich 1945-1955*, Vienna, Verl. f. Gesellschaftskritik.

Smith, A. D. (1992), 'National identity and the idea of European unity', *International Affairs*, 68 (1), 55-76 (p. 73).

Smith, D. (1989), *North and South: Britain's Economic, Social and Political Divide*, London, Penguin.

Spotts, F. & Wieser, T. (1986), *Italy, a Difficult Democracy*, Cambridge, Cambridge University Press.

Statistisches Jahrbuch für die Republik Österreich (1996), Vienna, Österreichisches Statistisches Zentralamt.

Sullivan, J. (1988), *ETA and Basque Nationalism: The Fight for Euskadi 1890-1986*, London, Routledge.

Svenska Kommunförbundet (1995a), *Developments in the European Union and their Effects on Sweden's Local Authorities*, Stockholm, Svenska Kommunförbundet.

Svenska Kommunförbundet (1995b), *Waves of Renewal: New Trends in Public Management and Administration in Swedish Local Authorities in the Nineties*, Stockholm, Svenska Kommunförbundet.

Syrett, S. (1995), 'Local Economic Development in Peripheral Rural Areas. The Case of Portugal', in Hardy, S., M. Hart *et alii* (eds), *An Enlarged Europe. Regions in Competition?* pp. 280-294.

Tálos, E. *et al* (eds) (1995), *Handbuch des politischen Systems Österreichs, Erste Republik 1918-1933*, Vienna, Manz.

Tamames, R. & T. Clegg (1984), 'Spain: regional autonomy and the democratic transition', in Hebbert, M. & H. Machin (eds), *Regionalism in France, Italy and Spain*, London, London School of Economics, International Centre for Economic and Related Disciplines.

Tarrow, S. (1977), *Between Center and Periphery: Grassroots Politicians in Italy and France*, New Haven, Yale University Press.

Telesis Report (1982), *A Review of Industrial Policy*, Dublin, National Economic and Social Council.

Terradas, I. (1989), 'Catalan Identities', in Herr, R. & J.H.R.Polt (eds), *Iberian Identity: Essays on the Nature of Identity in Portugal and Spain*, Berkeley, Institute of International Studies.

Theodorou, T. E. (1995), *I Ellininiki Topiki Autodiikisi; Tomos Tritos I Deuterovathmia Topiki Autodiikisi [Greek Local Government: Third Volume, Second Tier Local Government]*, Athens, Ekdosis Tolidi.

Thomas, A. H. (1996), 'The Concept of the Nordic Region and the Parameters of Nordic Cooperation', in Miles, L. (ed.) *The European Union and the Nordic Countries*, London, Routledge, pp. 15-31.

Thomas, P. (1990), 'Belgium's North-South Divide and the Walloon Regional Problem', *Geography*, 326, 75 (1) (January 1990), 36-50.

Tighe, C., 'Portsmouth to compete for EC aid', *Financial Times*, 12 October 1983.

Todd, E. (1990), *L'Invention de l'Europe*, Paris, Seuil.

Tommel I. (1992), 'Decentralisation of regional development policies in the Netherlands – A new type of state intervention?' in *West European Politics*, 15 (2), 107-125.

Toonen, T. (1993), 'Dutch provinces and the struggle for the meso', in L. Sharpe (ed.), *The Rise of Meso Government in Europe*, London.

Uhrich, R. (1987), *La France inverse, les régions en mutation*, Paris, Economica.

Valente, V. P. (1983), *Tentar Perceber*, Lisbon, Imprensa Nacional.

Valle, T. del (1989), 'Basque ethnic identity at a time of rapid change', in Herr, R. & J.H.R.Polt (eds), *Iberian Identity, Essays on the Nature of Identity in Portugal and Spain*, Berkeley, Institute of International Studies.

Vassallo, S. (1997), 'Il federalismo sedicente', *il Mulino*, 4 (July-August), 694-708.

Verney, S. (1994), 'Central State-Local Government Relations', in Kazakos & Ioakimidis, *Greece and EC Membership Evaluated*, pp.166-180.

Verney, S. & Papageorgiou, F. (1992), 'Prefecture Councils in Greece, Decentralisation in the European Community Context', *Regional Politics and Policy*, 2 (1), 109-138.

Vos, Louis (1993), 'Shifting nationalism, Belgians, Flemings and Walloons', in Teich, M. & R. Porter (eds), *The National Question in Europe in Historical Context*, Cambridge, Cambridge University Press.

Waever, O. (1993), 'Europe since 1945: crisis to renewal', in van der Dussen, J. & K. Wilson (eds), *The History of the Idea of Europe*, Milton Keynes, The Open University.

Wagstaff, P. (1994), *Regionalism in Europe*, Oxford, Intellect, 1994.

— (1996), 'Nations, Regions and the Future of Europe', *Journal of Area Studies*, 9, 126-141.

Walsh, J. (1995a), *Regions in Ireland, A Statistical Profile*, Dublin, Regional Studies Association (Irish Branch).

— (1995b) 'Economic Geography: How Ireland's Wealth is Dispersed', in D'Arcy & Dickson (eds), *Border Crossings*, 53-65.

Walsh, J. & D. Gillmor (1993), 'Rural Ireland and the Common Agricultural Policy', in King, R. (ed.), *Ireland, Europe and the Single Market*, pp. 84-100.

Weber, E. (1977), *Peasants into Frenchmen: the modernization of rural France 1870-1914*, London, Chatto & Windus.

Weber, F. (1995), 'Wiederaufbau zwischen Ost und West', in Sieder, R., H. Steinert & E. Tálos (eds), *Österreich 1945-1955: Gesellschaft, Politik, Kultur*, Vienna, Verl. f. Gesellschaftskritik, pp. 68-79.

Weinzierl, E. (1955), 'Zeitgeschichte im Überblick', in Dusek, Pelinka, Weinzierl, *Zeitgeschichte im Aufriß*, 4th edition, Vienna, Jugend & Volk.

Wendt, F. (1979), *Cooperation in the Nordic Countries,* Copenhagen, The Nordic Council.

White, M. (1993a), 'Confusion on urban renewal supremos', *The Guardian*, 15 October 1993.

— (1993b), 'Regions regain planning clout', *The Guardian*, 5 November 1993.

Wilke, D. & B. Schulte (1990), *Der Bundesrat. Entwicklung des föderalen Verfassungsorgans*, Darmstadt, Wissenschaftliche Buchgesellschaft.

Willman J. & T. Burt, 'Business help sought in urban revival plan', *Financial Times*, 1 November 1993.

Wils, L. (1993), 'Belgium on the Path to Equal Language Rights up to 1939', in *Ethnic Groups and Language Rights* (*Comparative Studies on Governments and Non-Dominant Ethnic Groups in Europe, 1850-1940*, Vol. III), Dartmouth, New York University Press.

Wintour, P., 'Whitehall seeks EC aid for depressed South-East', *The Guardian*, 12 October 1993.

Witsen, J. (1991), 'Five decades, five directors: the national physical planning agency 1941-1991 – A personal view', *Built Environment* 17 (1), 65.

Witte, E. *et al* (1984), *Le Bilinguisme en Belgique: le cas de Bruxelles*, Brussels, Editions de l'Université de Bruxelles.

Witte, E. (1992), 'Belgian Federalism: Towards Complexity and Asymmetry', *West European Politics*, 15 (4), 95-117.

Zeldin, T. (1977), *France 1848-1945*, Oxford, OUP.

Zöllner, E., & T. Schüssel (1955), *Das Werden Österreichs*, Vienna, Tosa.

Elm Bank Publications

Multimedia CALL: Theory and Practice *Keith Cameron (ed)*
Published 1998. VII + 307pp. 0-9502595-9-4. £24.99/$44.95

A Tragic Farce: The Fronde (1648–1653) *Wendy Gibson*
Published 1998. VII + 148pp. 0-9502595-8-6. £14.99/$24.95

Dialogues 1: Ricochets *Susan Bainbrigge (ed)*
Published 1998. 96 pp. 1-902454-00-6. £9.99/$17.95

Dialogues 2: Endings *Ann Amherst and Kate Astbury (eds)*
January 1999. 100 pp. 1-902454-01-4. £9.99/$ 17.95

Contemporary French Pronunciation *Aidan Coveney*
January 1999. 100 pp. 1-902454-02-2. £9.99/$17.95

***The Coach* and *The Triumph of the Lamb*: Marguerite de Navarre** *Hilda Dale*
February 1999. 150 pp. 1-902454-04-9. £15.99/$29.95

Francophone Voices *Kamal Salhi (ed)*
February 1999. 150 pp. 1-902454-03-0. £24.99/$ 44.95

Francophone Studies: Discourse and Multiplicity *Kamal Salhi (ed)*
March 1999. 250 pp. 1-902454-05-7. £29.99/$49.95

André Breton – The Power of Language *Ramona Fotiade (ed)*
July 1999. 250 pp. 1-902454-06-5. £29.99/$49.95

Matthew Arnold's 'Church of Brou' and other poems: a new look *G.A. Featherston*
September 1999. 200 pp. 1-902454-07-3. £24.99/$44.95

Robert Garnier: Les Juifves *Keith Cameron (ed)*
Published 1996. X + 85pp. 0-9502595-7-8. £5.00/$9.95

As Mighty As The Sword: A Study of the Writings of Charles de Gaulle *Alan Pedley*
Published 1996. VII + 226pp. 0-9502595-3-5. £24.99/$44.95

Jonquils: A Florilegium of Literary Translations *Keith Cameron and Martin Sorrell (eds)*
Published 1996. V + 96pp. 0-9502595-5 1. £9.99/$17.95

Variability in Spoken French: A Sociolinguistic Study of Interrogation and Negation *Aidan Coveney*
Published 1996. V + 271pp. 0-9502595-4-3. £24.99/$44.95

The Short Story: Structure and Statement *William J. Hunter (ed)*
Published 1996. IX + 198pp. 0-9502595-2-7. £19.99/$34.95

Translation: Here and There, Now and Then *Jane Taylor, Edith McMorran & Guy Leclercq (eds)*
Published 1996. VII + 185pp. 0-9502595-6-X. £24.99/$44.95

Elm Bank Publications are distributed by
Intellect, FAE, Earl Richards Road North, Exeter EX2 6AS, UK.
Tel/Fax 44 (0)1392 475110 elmbank@intellect-net.com

- All US orders should be made to the US distributor, ISBS. Call Toll free 1 800 944 6190.
- Further information on these books and how to order them is available on Intellect's website, **www.intellect-net.com**, or can be supplied on request.